EXS 81

Frontiers in Biosensorics II

Practical Applications

Edited by F. W. Scheller
F. Schubert
J. Fedrowitz

Birkhäuser Verlag
Basel · Boston · Berlin

Editors
Prof. Dr. F.W. Scheller
Institut für Biochemie
und Molekulare Physiologie
c/o Max-Delbrück-Centrum
für Molekulare Medizin
Robert-Rössle-Strasse 10
D-13122 Berlin

Dr. F. Schubert
Physikalisch-Technische Bundesanstalt
Abbestrasse 2-12
D-10587 Berlin

Dr. J. Fedrowitz
c/o Centrum für Hochschulentwicklung
PO Box 105
D- 33311 Gütersloh

Library of Congress Cataloging-in-Publication Data
A CIP catalogue record for this book is available from the library of Congress,
Washington D.C., USA

Deutsche Bibliothek Cataloging-in-Publication Data
EXS. – Basel; Boston; Berlin: Birkäuser.
Früher Schriftenreihe
Fortlaufende Beil. zu: Experientia
81. Frontiers in Biosensorics. 2. Practical applications. – 1997
Frontiers in Biosensorics / ed. by. F.W. Scheller ... – Basel;
Boston; Berlin: Birkhäuser.
(EXS; 81)
ISBN-13: 978-3-0348-9884-3 e-ISBN-13: 978-3-0348-9045-8
DOI: 10.1007/978-3-0348-9045-8
NE: Scheller, Frieder [Hrsg.]
2. Practical applications. – 1997
Practical applications. – Basel; Boston; Berlin: Birkäuser. 1997
(Frontiers in Biosensorics; 2) (EXS; 81)
ISBN-13: 978-3-0348-9884-3

Softcover reprint of the hardcover 1st edition 1997

Printed on acid-free paper produced from chlorine-free pulp. TCF ∞

ISBN-13: 978-3-0348-9884-3

9 8 7 6 5 4 3 2 1

Contents

Frontiers in Biosensorics II
Practical Applications
ed. by. F.W. Scheller, F. Schubert and J. Fedrowitz

Reflectometric interference spectroscopy for direct affinity sensing

A. Brecht and G. Gauglitz

Institut für Physikalische und Theoretische Chemie, Universität Tübingen, D-72076 Tübingen, Germany

Summary. Molecular recognition by non covalent interactions is of key importance not only in fundamental biochemistry, but also in affinity-based analytics. In typical affinity assays labelled compounds are used for detection of assay response. In contrast, the label-free detection of molecular interaction allows a more straigthforward approach to binding detection, simplified test schemes, and additional information about kinetical characteristics of the interaction. Optical techniques are particulary useful in direct affinity detection. One approach, based on white light interferometry is discussed in detail. This technique monitors the change in thickness of surface-bound layers of biological material by white light interference. Applications are given from quantitative detection of high molecular weight analytes, detection of low molecular weight analytes in a competitive test scheme, direct detection of low molecular weight analytes with immobilised receptors, investigation of interaction kinetics, and thermodynamic analysis of binding equilibrium. Finally, an outlook with respect to low-cost bioanalytical systems and high throughput screening applications is given, comparing various transducers and demonstrating advantages of label-free detection.

Introduction

Biosensors make use of specific molecular interaction between at least one biological structure and another analyte species, often also a biomolecule. The specificity of such interaction processes is mediated by structural complementary of the interacting compounds. Important examples are biochemical signal transduction (receptor/transmitter interaction), metabolic activity (catalytic activity of enzymes) and recognition of foreign structures (immunological response). Therefore biospecific interaction is a widespread phenomenon of high importance, not only in biosensor applications.

Biomolecular interaction can be easily monitored if a detectable species is produced or consumed during the interaction or if another detectable effect is caused in the biological system. Catalytic activity of enzymes is in many cases detected easily by monitoring the concentration of a reactant. Biochemical signal transduction causes subsequent effects that can be monitored, at least in sufficiently complex biological systems (tissue, whole cells, subcellular fractions). Biomolecular interaction without linked biochemical "reporter" systems is less prone to detection and monitoring. Detection, monitoring and quantification of the interaction in these cases is much more challenging and demanding.

Specific biomolecular interaction is often described as a “key-lock interaction” or as a process of “induced fit”. Both terms refer to the structural complementary between a binder molecule and a ligand. The relevant range of binding constants observed in biological systems runs from about 10^5/M up to 10^{10}/M and more. Some low affinity interactions may be of importance, where a reactant is present in high concentrations as in some major metabolic pathways or in certain kinds of signal transduction. However, the predominant range is medium to high affinity. The standard free energy of complex formation therefore is about 40 kJ per mole. This must be compared to the considerably high molecular weight of biological compounds resulting in only minor thermal effects. Also molecular interaction or recognition processes often do not cause major structural changes. Therefore, spectroscopic or fluorimetric approaches (e.g. quenching of intrinsic fluorophores) are quite limited in the investigation and transduction of biomolecular interaction.

Effects that can be exploited are differences in the dissusion coefficient for the free and the bound state of a reactant. Detection of this quantity can be performed in a homogeneous phase with concomitant advantages. The effects are most pronounced, if the molecular weight of the reactants is highly different. Fluorescence-based methods supply appropriate probes for this technique (Kost et al., 1989; Eigen and Rigler, 1994).

If one of the reactants can be immobilized at a solid phase, the binding of the free compound will lead to a change in the distribution of this compound between the solution and the solid phase. This is the basis for many affinity-based systems in research as well as analytical applications. In many cases it is feasible to use a labeled compound in these tests. In that case the distribution of the label between the solid phase and the solution is quantified. The most prominent examples are from the field of immunoassays (Price and Newman, 1991; Tijssen, 1985). If the use of a label is unfavorable or even impossible, effects associated with the binding of a reactant at the solid phase must be exploited. For that purpose one reactant is immobilized at the surface of an appropriate physico-chemical transducer. The binding of a ligand at a surface will lead to distinct physical effects. Among the basic quantities that can be detected are the mass loading of the transducer surface, changes in the dielectric properties of the transducer/solution interface due to different dielectric functions of biological compounds and aqueous solution (Liedberg, 1983; Lukosz, 1995; Brecht and Gauglitz, 1995), and changes in the thickness of the biological adlayer. The first quantity (mass) can be detected by quartz microbalances (Tom-Moy et al., 1995). The second quantity can be observed over a wide frequency range. Therefore impedance spectroscopy (Bataillard et al., 1988) as well as refractometry (Daniels et al., 1988) (at frequencies of visible light) have been used. The last quantity – changes in the thickness of a biological adlayer – will result in overall effects at or below the actual size of a reactant, which is a few nanometers for biological macro-molecules and even less for low molecular weight ligands. Techniques capable of resolving such small distances are scanning probe microscopes (Davies et al., 1994), which have not yet found broad application in this field, or optical interferometric techniques, which achieve a resolution far beyond one nanometer.

This paper deals with an optical transducer of the latter kind. It discusses an interferometric method capable of resolving the minor changes in the thickness of a layer of biological material caused by the binding of sub-monolayer quantities of ligand molecules. The results presented here were obtained with this one particular transducer. However, the results are typical and illustrative for the potential of direct approaches in the detection and investigation of biomolecular interaction.

Reflectometric Interference Spectroscopy (RIfS)

The above-mentioned method of interference spectroscopy uses wavelength-dependent modulations observed in the reflectance at thin transparent films. This pattern is due to the superposition of reflected beams, thus the method is called Reflectometric Interference Spectroscopy (RIfS). It is one of the optical detection principles that have been applied successfully to direct monitoring of affinity interactions (Brecht et al., 1992a).

The basisc effect is white light interference at a thin transparent film, its thickness is in the range of one micron. At both the interfaces of this film some fraction of the incident radiation is reflected. These two reflected

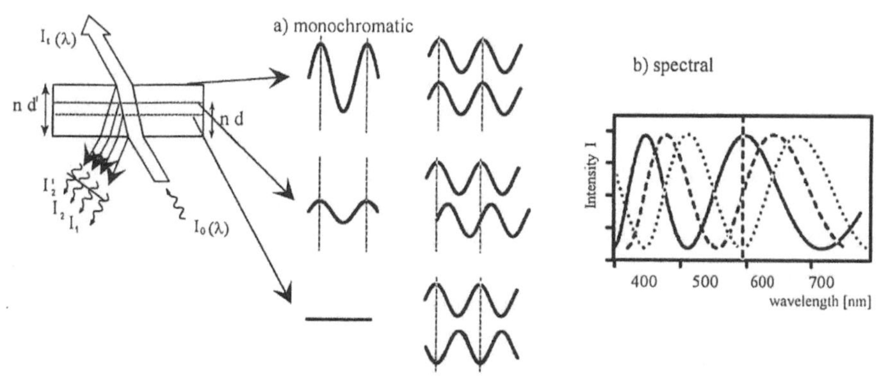

Figure 1. Principle of reflectometric interference spectroscopy (RIfS). Left: Reflection at the interfaces of a thin layer with three (assumed) different thicknesses. For graphical reasons the angle of incidence is chosen ≠ 0. Middle: At monochromatic radiation the two partially reflected superimposed beams show destructive or constructive interference or some amplitude in-between. Right: under polychromatic illumination the condition of interference varies with wavelength and an interference spectrum is obtained. Its extrema vary with the thickness of the layer.

partial beams can superimpose forming destructive or constructive interference (and some amplitude in-between) in dependence on the angle of incidence of radiation, its wavelength, the physical thickness of the layer, and its refractive index. If the reflectance at the interface is small (< 0.05) the treatment can be restricted to the first reflection at each of the interfaces. Thus, two reflected beams have to be considered (Gauglitz and Nahm, 1991). These two beams travel different paths and therefore a phase difference

$$\Delta\varphi = \frac{2nd}{\lambda} + \varphi_r \tag{1}$$

is introduced, n being the refractive index of the layer, d its physical thickness, λ the wavelength of incident radiation and φ_r any phase change upon reflection.

The principle is given in Figure 1. If the pathlength difference $2n \cdot d$ is less than the coherence length of the radiation, the two beams will interfere leading to a periodic modulation of the reflected light intensity according to

$$I = I_1 + I_2 + 2\sqrt{I_1 \cdot I_2 \cdot \cos\left(\frac{2\pi\Delta\varphi}{\lambda}\right)} \tag{2}$$

The most simple approach is to monitor the reflectance pattern under perpendicular incidence of light. Any increase in the thickness of the thin film will lead to a shift of the interference pattern to higher wavelengths. The absolute thickness of the layer can be derived from the reflectance pattern only after thorough characterization of the whole optical system. However, even minimum changes of the thickness of the layer can be quantified with high accuracy, just from the shift of the reflectance pattern.

Affinity reactions of biomolecules at a surface lead to the formation of biological layers that can be monitored and quantified in this way. The binding of analytes of high molecular weight leads to an increase in thickness of up to 10 nm per monolayer, while the binding of low molecular weight analytes will result in changes of only a few 100 pm per monolayer. A monolayer of biomolecules of about 10 nm thickness will not lead to a pronounced modulation of reflectance. To obtain a distinct interference pattern, the thickness of the layer has to be comparable to the wavelength of the radiation reflected. For this reason an optical interference layer is coated on a glass substrate. Preferably a material is chosen with a refractive index approximately the same as that of the biofilms to be observed. The binding of analytes during an affinity reaction can be monitored as an increase in the thickness of this interference layer. Changes in the thickness of such a layer can be measured with a resolution of the optical thickness ($n \cdot d$) down to 1 pm (Brecht and Gauglitz, 1994).

Set up

As the interference effect leads to a modulation of reflectance with wavelength, spectrometric techniques are the method of choice to observe binding effects. Especially photodiode array spectrometers allow fast aquisition of these reflection spectra. Any change in $\Delta\varphi$ (see eq. 1) will cause a change in the interference spectrum. The blunt end of an optical fiber is suitable for illumination and collection of reflected light, without a need for further optics or alignment.

The instrumental set-up is given in Figure 2. The flow cell (50 µm depth, approximately 150 nl volume) is connected to an ASIA FIA system from Isamtec (Zurich). A tungsten light source (20 W) is infrared filtered and coupled into one arm of a bifurcated fiber optics. The use of bifurcated (Y-shaped) fiber optics allows a flexible and simple optical setup. Various types of bifurcated fiber optics have been used successfully (sorted

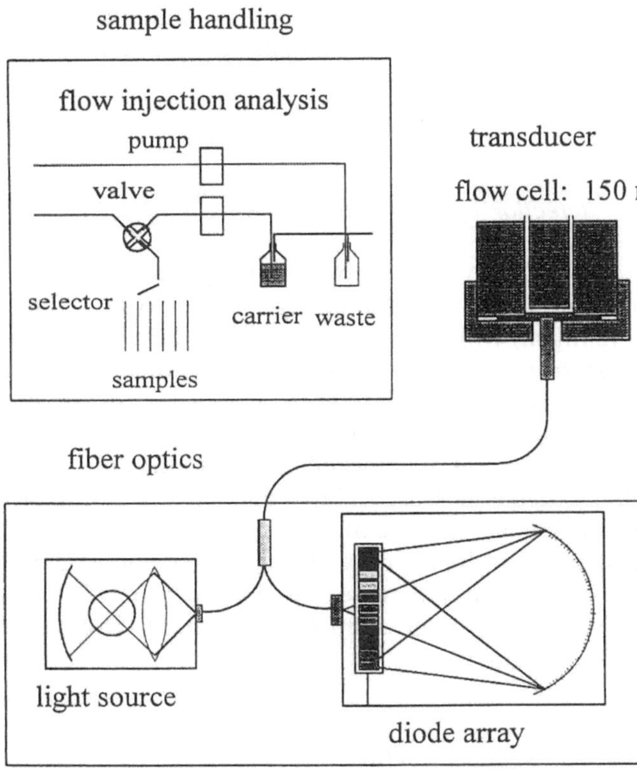

Figure 2. Set-up for RIfS measurements: the glass chip is mounted to a flow cell which is connected to a FIA-system; reflectance is monitored by a diode array spectrometer controlled by a computer.

bundles, random bundles, and true 2:1 fiber couplers). The common end of the fiber optics is placed perpendicularly to the interference layer on the uncoated side of the glass chip. A gap of approximately 100 μm between fiber and glass chip is ensured by a spacer to avoid interference effects. This gap between fiber optic and chip is filled with glycerol (80%) for index matching. The other free end of the fiber optic is coupled to the spectrometer. Data acquisition and control of the FIA system is performed by self-written software running under Windows on a personal computer.

A range of materials and deposition techniques has been investigated for the preparation of interference films (Brecht and Gauglitz, 1994). Materials tested were polymers, fluorides, and oxidic layers deposited by spin coating, sol-gel techniques, evaporation, and chemical vapor deposition. So far best results have been achieved with silica layers from a low temperature CVD process.

Also a range of strategies for the modification of the interference layer with one partner of the affinity system has been tested (Piehler et al., 1996b). Simple adsorption works well for many applications, however if low non-specific interaction and a potential for regeneration is required, best results have been achieved with a layer of aminodextran (Yalpani and Brocks, 1985), covalently linked to an interference layer made of silica.

Applications

Quantitative detection of high molecular weight analytes

The feasibility of RIfS in direct optical affinity sensing was demonstrated for the first time in an IgG and anti-IgG model system. The reaction of rabbit IgG (antigen) with goat-IgG (non-specific antibody) and goat-anti-rabbit-IgG (specific antibody), respectively, was monitored. A polystyrene film on glass served as the interference layer. After adsorption of the antigen to the substrate surface, non-specific interactions were blocked by ovalbumin. Incubation of the blocked layer with the non-specific antibody results in a negligible increase in layer thickness. Binding of the specific antibody resulted in an increase of thickness of about 2.5 nm. The changes in thickness on top of the interference layer are demonstrated in Figure 3. A calibration of thickness increase after a fixed time vs. antibody concentration is linear in the range from 0.5–25 μm/ml (Brecht et al., 1992a, 1992b, 1993) (see Fig. 4).

Kinetic measurements for interaction analysis

The high repetition frequency achievable with diode array detectors in spectral measurements allows to obtain time-resolved data of changes in

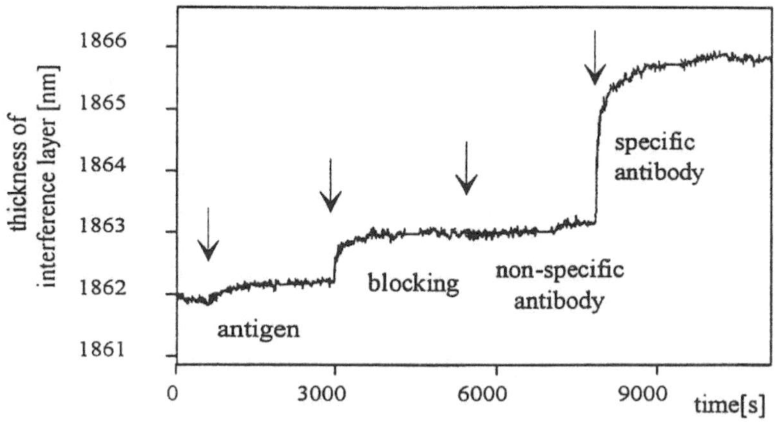

Figure 3. Increase in the thickness of the interference layer by different steps of bioreactions; the sequence investigated demonstrates the effects of rabbit-IgG (antigen) adsorption, blocking by ovalbumin, goat-IgG (non-specific antibody), and the reaction of rabbit-antigoat-IgG (specific antibody). The non-specific antibody exhibits a negligible change in thickness of 0.1 nm.

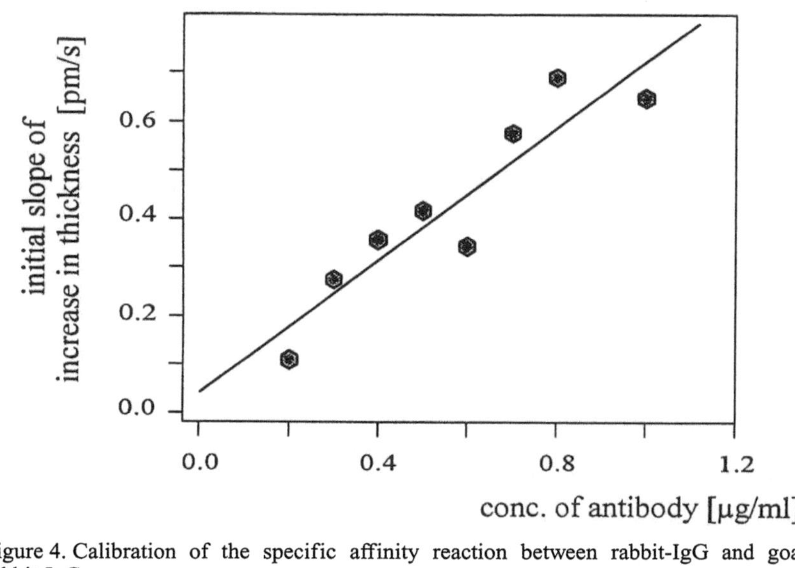

Figure 4. Calibration of the specific affinity reaction between rabbit-IgG and goat-anti-rabbit-IgG.

layer thickness. If binding reactions at the transducer surface are governed by reaction kinetics instead of mass transport, kinetical data can be derived from the binding curve, as discussed in (Eddowes, 1987; Fägerstam et al., 1992). Typically one reactant is immobilized and the concentration of the other compound is varied. The binding curves of such a set of measurements is graphed in Figure 5, while the secondary plots used to extract kinetical data are given in Figure 6.

Further evaluation is based on models for the binding process derived from homogeneous phase kinetics. A typical approach is to record binding curves at a range of concentrations and to derive the association rate constant from a set of linearized binding curves (Fig. 6). The dissociation rate constant is directly accessible from the off kinetics of a single binding curve, provided, that binding occurs only at a single binding site per molecule. The models applied are valid only for straightforward single-step reactions in the homogeneous phase. The rate constants obtained therefore are influenced by diffusion effects, by multiple binding events and the actual binding reaction taking place at the transducer surface. Therefore, the values determined by this method can differ from values determined in the homogeneous phase.

Quantitative detection of low molecular weight analytes (competitive assay)

Triazines are analytes of importance in environmental monitoring. Triazines are low molecular weight analytes that give only poor signals in a direct binding assay. Therefore, a competitive test protocol has been chosen for the quantification of triazines with the RIfS immunoprobe. A triazine derivative is immobilized to an aminodextran layer. In a pre-incubation step analyte solution (containing the different triazines) is mixed with the corresponding antibodies (Lang et al., 1996). After a short time equilibrium is reached and a fraction of the antibodies is occupied

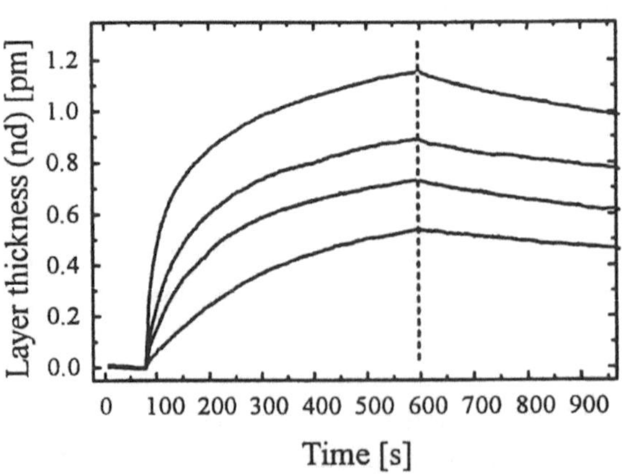

Figure 5. Time-resolved measurement of binding of an anti-pesticide antibody (c = 40 nM to 400 nM) to a pesticide derivative immobilized at the transducer surface. Left of dashed line: association kinetics; right of dashed line: dissociation kinetics.

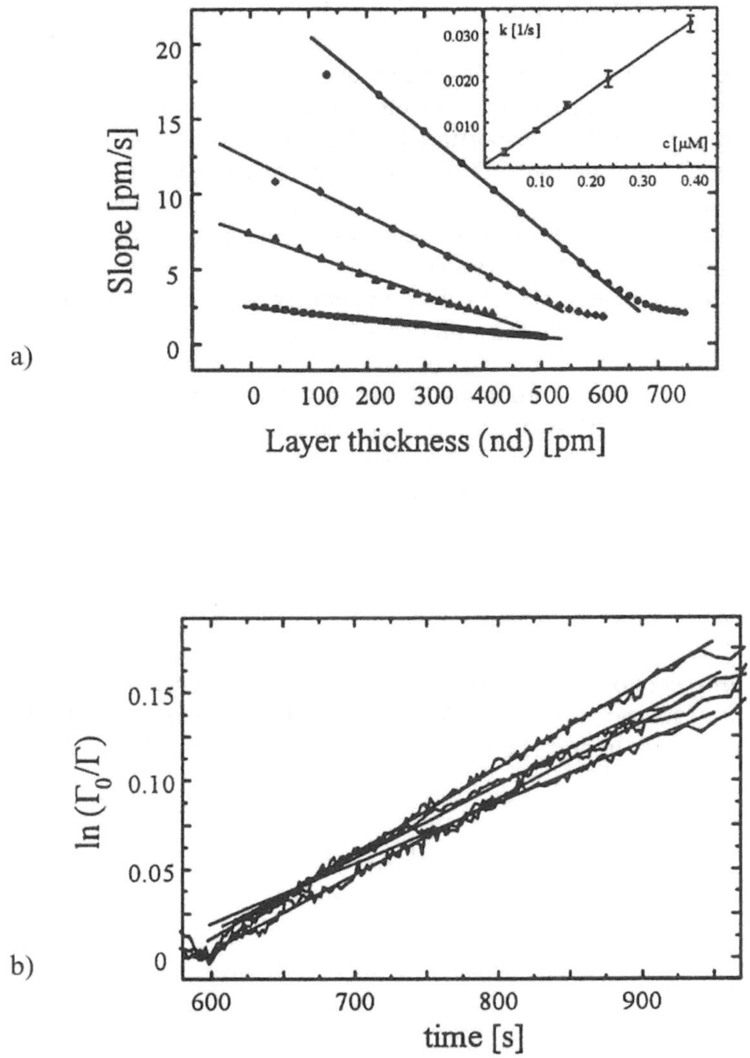

Figure 6. Secondary plots used to extract kinetical data from binding curves in Fig. 5 a) association rate constants are derived from the linearized binding curves of $d\Gamma$ vs Γ. Inset: plot of slope vs concentration to derive k_{ass}. b) dissociation rate constants are derived directly from the linearized off kinetics $\ln(\Gamma_0/\Gamma)$ vs t of the binding curve. Only little influence of the antibody concentration is observed.

(deactivated) by binding of free triazine. Subsequently the mixture is delivered to the transducer and unoccupied antibodies react with the triazine derivative immobilized at the transducer. Low analyte concentrations give a high antibody binding signal, while the opposite holds for high analyte concentrations.

A high coverage of the transducer with immobilized triazines leads to a fast binding reaction at the surface and hence the diffusion of antibodies to the surface becomes rate limiting. Therefore, under these circumstances the affinity reaction will be diffusion controlled. Then, linear binding curves are obtained, the slope being proportional to the concentration of free antibodies (see Fig. 7, time domain A). The midpoint of tests depends on the binding constant and the antibody concentration chosen (Brecht et al., 1995). A slight improvement can be achieved by using monovalent F_{ab} fragments of antibodies (Lang et al., 1996).

The immobilization of the stable low molecular weight compound in high density allows to remove the antibodies bound and thereby to regenerate the transducer surface. In the case of triazine measurement the sensitive layer could be regenerated more than 300 times with a loss in efficiency of no more than 5%. Low pH-values, chaotropic agents or Pepsin have been used to remove the antibodies from the surface. Afterwards the transducer is ready for a new cycle. The combined measurement and regeneration step is given in Figure 7.

Thermodynamical analysis of titration curves

In the preceding section a competitive test scheme was described for the quantification of unknown levels of low molecular weight analytes. If the low molecular weight ligand is added in known concentrations, the result-

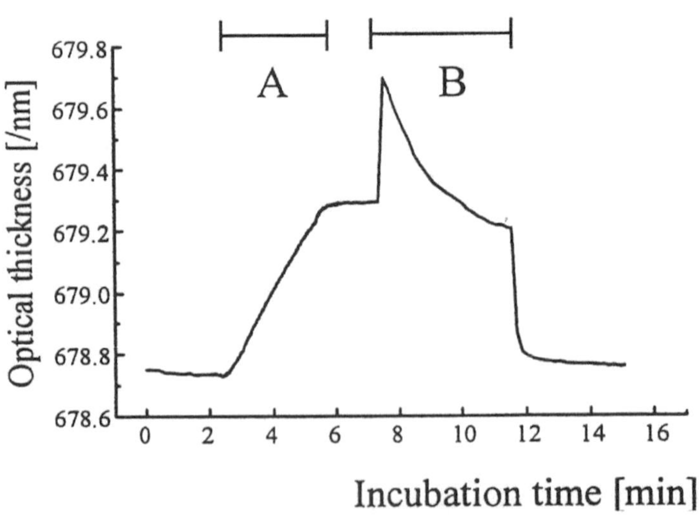

Figure 7. Test cycles of triazine immuno reaction and consecutive regeneration step; protocol: (A) injection of the preincubation solution containing antibody and triazine, (B) injection of pepsin and regeneration.

ing titration curve can be used to obtain information about the thermodynamics of the affinity reaction in the homogeneous phase (Piehler et al., 1995). The concentration of free binding sites is titrated from maximum response down to total inhibition of antibody binding. Antibody concentrations in the range of the reciprocal of the affinity constant are used. The evaluation approach must take into account the multivalency of binder molecules. An atrazin derivative immobilized to a dextran hydrogel. Atrazin was used as a model antigen in solution. The concentration of binding antibodies $c_{ab,bind}$, that is antibodies with unoccupied binding sites was assessed by RIfS. For bivalent antibodies the value of $c_{ab,bind}$ is given as a function of the initial concentrations of antibody ($c_{0,ab}$) and analyte ($c_{0,ag}$) and the equilibrium binding constant K by eq. 3:

$$c_{ab,bind} = \frac{c_{0,ab}}{2} - \frac{\left[\frac{c_{0,ab}+c_{0,ag}+\frac{1}{K}}{2} - \sqrt{\frac{\left(c_{0,ab}+c_{0,ag}+\frac{1}{K}\right)^2}{4} - c_{0,ab}\cdot c_{0,ag}}\right]^2}{2c_{0,ab}} \tag{3}$$

Titration models for monovalent F_{ab}-fragments and for bivalent antibodies are established. The degree of coverage of binding sites achieved during the pre-incubation phase is determined indirectly via the quantification of antibodies with free binding sites by RIfS and binding to the immobilized antigen. The affinity constant (free antigen-antibody) can be extracted from the titration curve measured. This model works for antibodies with high affinity constants (K1F4 and terbutylazine; K approx. 10^7 M^{-1}) in Fig. 8. Lower affinity constants are related with a higher value of k_{diss}. At high values of k_{diss} the dissociation of the antigen-antibody complex during exposition to the sensor leads to the formation of additional free antibody and an erroneous signal. For such low affinity interactions the equilibrium based model has to be extended (Piehler, 1996c).

Direct detection of low molecular weight analytes

To date, most work with direct affinity transducers has been directed to the detection of macromolecules with considerable molecular weight (>10 000 g/mol). In receptor-ligand systems with a high and a low molecular weight compound, typically the low molecular weight compound is immobilized for direct binding assays. However, in many applications the direct monitoring of binding of the low molecular weight is desirable. The monolayer signal for a given compound will approximately increase with the cubic root of the molecular weight. For a thousandfold decrease of the molecular weight this leads to an acceptable decrease in maximum signal by a factor of ten. However, in binding assays with a high molecular weight

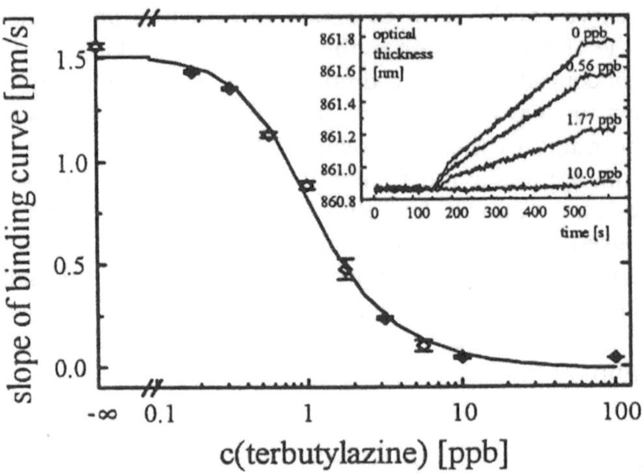

Figure 8. Triation of antibody K1F4 (0.75 μg/ml) with terbutylazine and evaluation of binding curves by fit to equilibrium binding model. Inset: some of the raw data (binding curves) recorded for different concentrations of analyte.

compound at the transducer surface, the surface coverage for the recpetor compound limits the maximum binding effect. Therefore, in direct binding assays with low molecular weight compounds, signal levels tend to be two to three orders of magnitude below the values observed with high molecular weight compounds.

A first approach in that field was based on surface plasmon resonance (Liedberg et al., 1983) where the detection of low molecular weight analytes (biotin at a concentration of 400 μM) has been demonstrated recently (Karlsson and Stahlberg, 1995). We have undertaken a similar set of experiments with the RIfS transducer where the binding between streptavidin immobilized via biotinylated proteins to the sensor surface and biotin in the analyte solution (Piehler et al., 1996a) was examined. The immobilization protocol used results in some baseline drift. However, the thickness change due to biotin interaction could be clearly identified down to a concentration of 40 nM biotin. A linear relationship between the amount of streptavidin immobilized and the signal change during incubation with biotin was found. To check the increase of the thickness for non-specificity benzoic acid was injected as another organic compound with a free carboxylic acid group. Even at rather high benzoic acid concentrations (40 μM) no binding effect was observed. This proves the biotin binding effects are due to specific molecular interaction.

The given test protocol allows to monitor binding effects between immobilized receptor compound of high molecular weight and a low molecular weight ligand. This is of particular importance in the investigation of biomolecular interaction and recognition processes. The data prove that a surface coverage of a few picograms/mm^2 of organic matter can be detected.

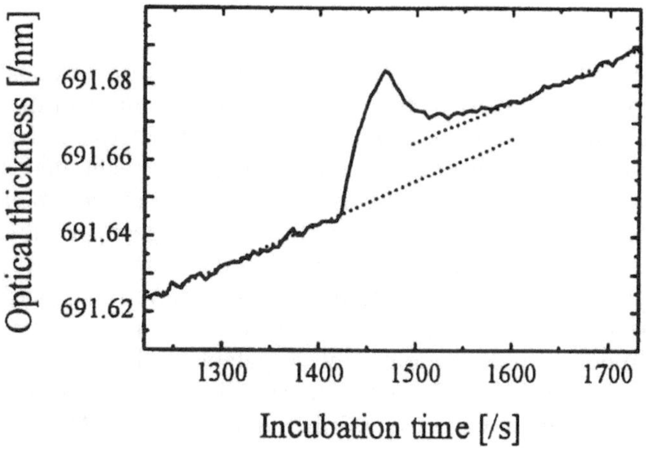

Figure 9. Binding ob biotin (40 nM) at poly-streptavidin surface, which amounts to 10 pm change in thickness.

Table 1. Limit of detection of various direct optical transducer principles.

Sensing principle	System	Limit of detection	Reference
Surface Plasmon Resonance	BIACore (Pharmacia)	5–0.3 pg/mm^2	(Karlsson and Stahlberg, 1995)
Frustrated total reflection	Resonant Mirror (Fisons)	5 pg/mm^2	(Edwards et al., 1995)
Grating coupler	Grating coupler (ASI)	10–5 pg/mm^2	(Lukosz et al., 1991)
Differential mode interferometry		2 pg/mm^2	(Fattinger et al., 1993)
Dual beam waveguide interferometer		few pg/mm^2	(Heideman et al., 1993)
Waveguide surface plason resonance		2 pg/mm^2	(Harris and Wilkinson, 1995) Harris, personal communication
Reflectometric Interference Spectroscopy (RIfS)		5–1 pg/mm^2	(Brecht et al., 1994)

Outlook

A variety of optical transducers has been applied to direct optical sensing of immunoreactions. Most of them have a detection limit in the range of 1 – 10 pg/mm^2 biocomponents either observed by refractive index changes via the evanescent field (micro refractometers (Brecht and Gauglitz, 1995)) or as given here in detail by use of white light reflectometry. Direct optical detection of low molecular weight analytes seems possible.

The potential use of transducers for biomolecular interaction will be broadened by the availability of transducer arrays. A transducer array in that case means a multitude of distinct transducer elements in close vicinity or a multitude of independently interrogatable active areas on the surface of a single transducer element. An array therefore allows to bring the same sample in contact with a set of (differently pre-treated) surfaces.

The major benefit in the field of interaction analysis (applied or basic research) will be the possiblility to compare the effects between a specific surface and a reference surface. This allows a more effective suppression of non-specific binding effects. This becomes more important, if transient effects in front of a large background are to be monitored.

In the field of "real" biosensing, many applications do not require the detection of a single analyte, but a multitude of substances. Transducer arrays would allow the effective detection of a multitude of analytes from a single sample without an increase in sample handling and pretreatment procedures. Arrays with up to 10 individual positions would already serve a wide range of applications from the biomedical field, from process control, from food testing, or even from environmental monitoring.

An emerging field where the label free detection of molecular interaction may become an important tool is the testing of tentative drugs in pharmaceutical screening and the search for new leads. This field is stimulated by the increasing knowledge about key compounds in metabolism which is provided by molecular biology. These molecular targets become more and more available from recombinant organisms. The potential of automated and combinatorial synthesis gives access to millions of new structures.

In most cases pharmaceutical effects can be traced back to an initial molecular recognition event. Therefore the detection of binding effects opens a very generic route to screening for substances with potential pharmacological activity. The use of label-free approaches is especially attractive as changes of biological activity due to the introduction of a label can be ruled out.

The prime requirement for molecular screening systems based on direct bioaffinity transducers is a high throughput. For the heterogeneous (solid phase based) type of transducer discussed here, diffusion of reactants dictates the necessary time per cycle. Typical test cycles are in the range of a few minutes (depending on concentrations used). To meet throughput requirements in the range of a million samples per year a parallel arrangement of transducers is required. Therefore, also this application requires arrayed transducer structures. A system capable of handling large amounts of samples and interfacing to the well-estabished microtiter plate format would be desirable. This would mean about 100 or up to about 400 channels (with emerging plate formats) per system. If this throughput could be combined with high sensitivity systems that allow the detection of low molecular weight analytes in direct binding assays, this could well open a new field of applications.

Conclusion

Direct detection is highly attractive for basic and applied research (interaction analysis). Also some potential in the field of affinity-based analytical procedures can be demonstrated. In analytical applications the well-established immunoassays are a strong competitor because of market aspects (well-introduced techniques as ELISA, RIA etc.), and because of detection limits clearly below the present range of direct optical transducers.

Pharmaceutical drug screening may become a field, where the particular advantages of label-free approaches are especially attractive. However, not all of the direct optical transducers will allow to be paralleled.

As could be demonstrated, the emerging level of sensitivity allows to monitor small molecules binding. Systems with increased throughput and decreasing limits of detection lead the way to an exciting field for further research and new analytical applications.

References

Bataillard, P., Gardies, F., Jaffrezic-Renault, N., Martelet, C., Colin, B. and Mandrand, B. (1988) Direct Detection of Immunospecies by Capacitance Measurements. *Anal. Chem.* 60:2374–2379.

Brecht, A. and Gauglitz, G. (1994) Optimised layer systems for immunosensors based on the RIfS transducer. *Fresenius J. Anal. Chem.* 349:360–366.

Brecht, A. and Gauglitz, G. (1995) Optical probes and transducers. *Biosens. Bioelectron.* 10:923–936.

Brecht, A., Kraus, G. and Gauglitz, G. (1992 a) Interferometric measurements used in chemical and biochemical sensors. *Analusis* 20:135–140.

Brecht, A., Ingenhoff, J. and Gauglitz, G. (1992 b) Direct monitoring of antigen-antibody interactions by spectral interferometry. *Sensor. Actuator. B* 6:96–100.

Brecht, A., Gauglitz, G. and Polster, J. (1993) Interferometric immunoassay in a FIA-system: a sensitive and rapid approach in label-free immunosensing. *Biosens. Bioelectron.* 8: 387–392.

Brecht, A., Piehler, J., Lang, G. and Gauglitz, G. (1995) A direct optical immunosensor for atrazine detection. *Anal. Chim. Acta* 311:289–299.

Daniels, P.B., Deacon, J.K., Eddowes, M.J. and Pedley, D.G. (1988) Surface Plasmon Resonance Applied to Immunosensing. *Sensor. Actuator.* 15:11–18.

Davies, J., Roberts, C.J., Dawkes, A.C., Sefton, J., Edwards, J.C., Glasbey, T.O., Haymes, A.G., Davies, M.C., Jackson, D.E., Lomas, M., Shakesheff, K.M., Tendler, S.J.B., Wilkins, M.J., Williams, P. M. (1994) Use of Scanning Probe Microscopy and Surface Plasmon Resonance as Analytical Tools in Study of Antibody-Coated Microtiter Wells. *Langmuir* 10: 2654–2661.

Eddowes, M.J. (1987/88). Direct immunochemical sensing: basic principles and fundamental limitations. *Biosensors* 3:1–15.

Edwards, P.R., Gill., A., Pollard-Knight, D.V., Hoare, M., Buckel, P.E., Lowe, P.A. and Leatherbarrow, R.J. (1995) Kinetics of protein-protein interactions at the surface of an optical biosensor. *Anal. Biochem.* 231:210–217.

Eigen, M. and Rigler, R. (1994) Sorting single molecules: Application to diagnostics and evolutionary biotechnology. *Proc. Natl. Acad. Sci. USA* 91:5740–5747.

Fägerstam, L., Frostell-Karlsson, A., Karlsson, R., Persson, B. and Rönnberg, I. (1992) Biospecific interaction analysis using surface plasmon resonance detection applied to kinetic, binding site and concentration analysis. *J. Chromatogr.* 597:397–410.

Fattinger, Ch., Koller, H., Schlatter, D. and Wehrli, P. (1993) The difference interferometer: a highly sensitive optical probe for quantification of molecular surface concentration. *Biosens. Bioelectron.* 8:99–107.

Gauglitz, G. and Nahm, W. (1991) Observation of spectral interferences for the determination of volume and surface effects of thin films. *Fresenius J. Anal. Chem.* 341:279–283.

Harris, R.D. and Wilkinson, J.S. (1995) Waveguide surface plasmon resonance sensors. *Sensor. Actuator.* *B*29:261–267.

Heideman, R.G., Kooyman, R.P.H. and Greve, J. (1993) Performance of a highly sensitive optical waveguide Mach-Zehnder interferometer immunosensor. *Sensor. Actuator.* *B*10:209–217.

Karlsson, R. and Stahlberg, R. (1995) Surface plasmon resonance detection and multispot sensing for direct monitoring of interactions involving low-molecular-weight analytes and for determination of low affinities. *Anal. Biochem.* 228:274–280.

Kost, G.J., Vogelsang, Ph.J., Reeder, B.L., Omand, K.L. and Leach, C.S. (1989) Future trends in automation of nonisotopic immunoassay. *Lab. Robotics Automation* 1:275–284.

Lang, G., Brecht, A. and Gauglitz, G. (1996) Characterisation and optimisation of an immunoprobe for triazines. *Fresenius J. Anal. Chem.* 354:857–860.

Liedberg, B., Nylander, C. and Lundström, I. (1983) Surface plasmon resonance for gas detection and biosensing. *Sensor. Actuator.* 4:299–304.

Lukosz, W. (1995) Integrated optical chemical and direct biochemical sensors. *Sensor. Actuator.* *B*29:37–50.

Lukosz, W., Clerc, D. and Nellen, Ph.M. (1991) Input and Output Grating Couplers as Integrated Optical Biosensors. *Sensor. Actuator.* *A*25:181–184.

Piehler, J. (1996) PhD-Thesis, Tübingen, in press.

Piehler, J., Brecht, A., Kramer, K., Hock, B. and Gauglitz, G. (1995) Multi-analyte determination with a direct optical multi-antibdy detection system. *Proc. SPIE* 2504:185–194.

Piehler, J., Brecht, A. and Gauglitz, G. (1996a) Affinity Detection of Low Molecular Weight Analytes. *Anal. Chem.* 68:139–143.

Piehler, J., Brecht, A., Geckeler, K.E. and Gauglitz, G. (1996b) Surface Modification for Direct Immunoprobes. *Biosens. Bioelectron.* 11:579–590.

Price, Ch.P. and Newman, D.J. (eds) (1991) *Principles and practice of immunoassay*. Stockton Press, New York.

Tijssen, P. (1985) *Practice & theory of enzyme immunoassay.* Elsevier Science Publisher, Amsterdam.

Tom-Moy, M., Baer, R.L., Spira-Solomon, D. and Doherty, Th.P. (1995) Atrazine measurements using surface transverse wave devices. *Anal. Chem.* 67:1510–1516.

Yalpani, M. and Brooks, D.E. (1985) Selective chemical modifications of dextran. *J. Polym. Sci.* 23:1395–1405.

Frontiers in Biosensorics II
Practical Applications
ed. by. F.W. Scheller, F. Schubert and J. Fedrowitz

Immunosensors for clinical diagnostics

C.J. McNeil[1], D. Athey[2] and R. Renneberg[3]

[1]*Department of Clinical Biochemistry, Medical School, University of Newcastle upon Tyne, NE2 4HH, U.K.;*
[2]*Cambridge Life Sciences plc, Cambridgeshire Business Park, Angel Drove, Ely, CB7 4DT, U.K.;*
[3]*Department of Chemistry, The Hong Kong University of Science and Technology, Kowloon, Hong Kong.*

Overview

Enzmye immunoassays couple the amplification produced by enzymatic reactions to the sensitivity of the immune reaction. The growing trend away from radioimmunoassay, and the increasing number of enzyme-linked immunoassays, coupled with the wide range and low detection limits of the electroanalytical methods, has resulted in a proliferation of papers linking immunoassays to electrochemical means of detection. This review will concentrate on electrochemical (mainly amperometric) approaches to the design of enzyme immunosensors for clinical diagnostics. It should be recognised that many of the approaches could be easily adapted to other areas where immunodiagnostics have an essential role, for example in environmental monitoring and the food industry.

The concept of using electrochemistry to monitor immunological reactions is now new. Breyer and Radcliffe (1951) used polarography to quantitate an antigen-antibody reaction. Azo-protein was used as an antigen, and the reaction of the azo-protein with specific and non-specific antisera was monitored. However, it was some time before the development of modern electronics and electrochemistry allowed the technique to develop into the powerful method that it is today. The rationale behind the development of electrochemical immunosensor systems is that such systems should be sensitive due to the characteristics of the electrochemical detection methods, whilst exhibiting specificity due to the antigen-antibody reaction (Heineman and Halsall, 1985; Jenkins et al., 1991).

Amperometric-based enzyme immunosensor approaches can be divided conveniently as follows:

a) Assays based on the Clark oxygen electrode. These employ enzymes that can either consume or produce oxygen in the presence of suitable substrates.
b) Amperometric assays which use an enzyme label and detect electrochemically products of that enzyme.

Amperometric immunosensors based on the Clark oxygen electrode

These electrochemical immunoassays have used the Clark oxygen electrode to detect either the consumption or formation of oxygen as a consequence of an enzyme reaction. The two most commonly used labels have been glucose oxidase (consumption of O_2) and catalase (production of O_2).

Aizawa et al. (1979) constructed an enzyme-linked immunoassay to monitor human chorionic gonadotrophin (hCG) using catalase labelled hCG. The antibody was immobilised onto a pre-cast cellulose membrane, the membrane was then placed over the Teflon membrane of the oxygen electrode. Both labelled and non-labelled hCG competed for the membrane-bound antibody. The membrane was then washed to remove unbound hCG, and the electrode exposed to hydrogen peroxide solution. Catalase converted hydrogen peroxide to oxygen and water. The rate of increase in oxygen tension was monitored.

The other alternative in this approach is to concentrate on the consumption of oxygen rather than its production. A glucose oxidase electrode has been used in an assay involving the enzyme label alkaline phosphatase (Renneberg et al., 1983). This involved using the substrate glucose-6-phosphate with alkaline phosphatase, and then converting the glucose to H_2O_2 with glucose oxidase. The assay was complicated in that it needed two enzymes, the reaction time was slow and the assay was not sensitive.

Enzyme-labelled amperometric immunoassay

Enzyme labels have become the most widely used labels for immunoassay methodology owing to their inherent amplification feature, which allows lower concentrations of analyte to be detected by increasing the substrate incubation time used in the assay. They also allow for a wide variety of detection schemes depending on the choice of enzyme and substrate. The following are the commonest labels used for this purpose in amperometric electrochemical immunoassay.

(a) Alkaline phosphatase (ALP)

Alkaline phosphatase and catalyse the hydrolysis of phosphate esters to give inorganic phosphate and a phenolic leaving group. The formation of these phenolic moieties has been followed by numerous spectroscopic methods. Electrochemical detection for a number of phosphatase substrates has been described and, historically, the bulk of this literature concerns the use of phenyl phosphate (Heineman and Halsall, 1985). Phenol can be oxidised at +750 mV Vs Ag/AgCl in carbonate buffer, whereas the substrate phenyl phosphate is electro-inactive at positive potentials. However, electrochemical monitoring of phenol is complicated by the fouling of the electrode surface due to the electropolymerisation of the phenoxy radical formed in the one electron oxidation of phenol. Further difficulties may arise when using biological samples due to protein adsorption at the electrode surface. This phenomenon can result in reduction of the current response. Problems such as these stimultaed the design of electrochemical immunoassays in which chromatographic separation was combined with amperometric detection of the enzyme-generated product. These systems attempted to address and overcome problems such as electrode fouling and generally aimed to increase assay sensitivity.

Some of the problems associated with detection *via* direct oxidation of phenol have been addressed in a recent publication from Scheller's group who have very effectively combined the two amperometric-based enzyme immunosensor approaches outlined previously (Bauer et al., 1996). These authors developed a "phenol-indicating" sensor which consisted of a Clark oxygen electrode covered by a membrane containing coentrapped tyrosinase and quinoprotein glucose dehydrogenase. Phenol, formed by the reaction of an alkaline phosphatase label with phenylphosphate, was oxidised in the membrane by tyrosinase to form ortho-quinone (*via* catechol) with concomitant oxygen consumption. The quinone was then re-converted to catechol by glucose dehydrogenase and the overall enzyme cycling scheme resulted in a 350-fold amplification of the sensor response to phenol. This system could detect 3.2 fM ALP.

Recently, the use of 4-aminophenyl phosphate (4-APP) for amperometric detection of ALP activity has found considerable favour, due to the reversible electrochemistry of the product 4-amino phenol (4-AP) at + 100 mV vs Ab/AgCl. This substrate was first synthesised in 1957 (Boyland and Manson, 1957), but was not recognised as an improved substrate for ALP until much later (Kulys et al., 1980). The first use of 4-APP as a substrate in an electrochemical immunoassay was attributed to Tang and co-workers (Tang et al., 1988). Since that time there has been increased interest in the substrate. Electrochemical immunoassays for human chorionic gonadotrophin (Rosen, and Rishpon, 1989), phenytoin (Frew et al., 1989), theophylline (Gil et al., 1990), apoliprotein-E (Meusel et al., 1995) and fatty acid binding protein, an early marker for acute myocardial infarction (Glatz et al., 1995; van Nieuwenhoven et al., 1995; Siegmann-Thoss et al., 1996), using 4-APP have been reported. The product is oxidised at an electrode to a quinone-imine (4-quinone-imine). Thompson et al. (1991) compared amperometric determination of 4-APP with conventional spectrophotometric assays, based on 4-nitrophenyl phosphate. The detection limit using 4-APP with flow injection alanysis was twenty times lower than that for spectrometric determination involving 4-nitrophenyl phosphate.

In addition to the systems described using 4-APP, alternative phosphatase substrates have been employed for the development of electrochemical immunosensor instrumentation. For instance, Athey et al. (1993 a) proposed the use of 1-naphthyl phosphate in conjunction with disposable scren printed carbon electrodes for the development of a lap-top sized electrochemical immunoassay analyser which was manufactured by Immunosens SpA, Benevento, Italy. The choice of this substrate was based on the relatively low oxidation potential employed using screen printed carbon electrodes, the fact that the substrate was available commercially and that the product was stable at alkaline pH, unlike 4-AP. This instrument was used in clinical assays of thyroid stimulating hormone (Athey et al., 1993). The system was developed further to introduce a generic approach to immuno-electrode development by immobilisation of antibodies on screen printed avidin-coated disposable electrodes (Athey, 1993).

(b) Glucose oxidase (GOD)

The detection of GOD labels is not restricted to the use of the Clark oxygen electrode. Other methods include the direct oxidation of the hydrogen peroxide produced in the enzymatic reaction. For instance, deAlwis and Wilson (1985) described a two-site ELISA for IgG involving the use of the high performance chromatography. The second antibody was glucose oxidase-goat anti IgG conjugate. The affinity column contained immobilised bovine IgG and peroxide in the column effluent was detected following injection of glucose.

Direct electron transfer between enzymes and electrodes is difficult to achieve; at best it requires the careful choice of electrode surface and solution conditions. The alternative is to use small, electroactive species (mediators) to enhance the rate of electron transfer. In the case of glucose oxidase the requirement is for a molecule that can replace oxygen as the electron acceptor. Ferrocene has proved particularly useful in this role (Cass et al., 1984). The ferricinium ion can be regenerated by oxidation at an electrode.

$$2\,Fecp_2R \rightarrow 2\,Fecp_2R^+ + 2\,e^- \rightarrow \text{electrode current}$$

It has been shown that when the flavin within glucose oxidase is reduced in the presence of glucose, the oxidised form of the enzyme may be regenerated by electron transfer to the ferricinium ion, generated at the electrode. The system is therefore not dependant on oxygen as the ferricinium ion becomes the co-factor for glucose oxidase. This reaction scheme forms the basis of an amperometric immunoelectrode developed by diGleria et al. (1986) using lidocaine as the model analyte. The assay involves the use of a ferrocene-drug conjugate as an electron acceptor for glucose oxidase. On binding of the conjugate to its specific antibody the electron transfer properties are lost. Addition of free drug displaces some conjugate, decreasing the inhibition, the difference in the currents reflects the amount of the drug present.

A rapid two-site immunoassay using an electrode-immobilised capture antibody has been reported for the measurement of hCG (Robinson et al., 1986). The activity of glucose oxidase bound to the second antibody was determined electrochemically by the use of a ferrocene electron transfer mediator without the need for an incubation step. The mean assay sensitivity was 9 IU/L hCG, the levels of hCG in healthy men and healthy non-pregnant women are normally below 10 IU/L. This type of homogeneous electrochemical immunoassay has also been demonstrated for theophylline (Hill, 1984).

(c) Glucose 6 phosphate dehydrogenase

Glucose 6 phosphate dehydrogenase in common with over 250 other dehydrogenase enzymes has a requirement for the co-factor nicotinamide adenine dinucleotide (NADH). Considerable effort has been directed towards designing electrochemical immunoassays which use G6PDH as an enzyme label. NADH can be oxidised via a single step two-electron and one-proton loss with the nicotinamide moiety as the electroactive centre (Bresnahan et al., 1981; Elving et al., 1982). However, the difficulty in effecting NADH oxidation at an electrode is well known, and a considerable body of work has been devoted to the development of catalytic surfaces which will allow this process to occur at low potentials (Bartlett and Whittaker, 1988).

The most successful homogeneous enzyme immunoassay over recent years has been the EMIT assay (Rubenstein et al., 1972), a spectrophotometric assay that measures reduction of the co-factor NAD^+ as the assay signal. Attempts have been made to convert this type of assay to electrochemical detection. Phenytoin, an anti-epileptic drug, has been measured using the EMIT format, in combination with flow injection analysis (FIA) linked to amperometric detection of NADH (Eggers et al., 1982). The optimum detection range for NADH was at one hundredth of the concentration of NADH generated during the period required for the analysis. An approximately 100-fold dilution of the sample was therefore required immediately prior to injection into the flow system.

Broyles and Rechnitz (1986) described an assay for drug antibodies using amperometric detection based on the inhibition of the enzyme activity of an enzyme-antibody conjugate by the antibody that is to be measured. Enzyme activity was monitored by the amperometric determination of the rate of NADH oxidation at a platinum electrode. The technique used the lidocaine/anti-lidocaine system as a model and dose response curves were obtained at nanogram levels of antibody. The absolute sensitivity of the assay was dependent on the original amount of enzyme conjugate. Interferences such as contamination of the platinum electrode were shown to have minimal effect.

More recently, McNeil's group (Athey et al., 1993b; Manning et al., 1994) have applied platinised activated carbon electrodes (PACE) to electrochemical EMIT assays for theophylline and digoxin. PACE allows the highly reproducible and sensitive measurement of NADH oxidation at low overpotential (+150 mV vs Ag/AgCl) and facilitated the development of systems capable of measurement in whole blood.

Recent advances in electrochemical immunosensors

The analytical advantages of immunoassays based on non-competitive approaches stem from the removal of unbound, or non-specifically bound materials (excess analyte, excess enzyme-labeleld antibody etc.) from the system prior to measurement of the activity of the bound enzyme label. Sandwich-type immunoassays thus exhibit greater sensitivity and specificity, however, they require numerous washing steps and are not ideally suited for use in decentralised diagnostics (e.g. doctor's office, emergency room and coronary care unit). Therefore a primary goal of clinical immunosensor technology has been the development of simple, rapid methods for separation-free non-competitive measurement in whole blood.

Duan and Meyerhoff (1994) described an extremely elegant approach to such a potential system. These authors employed gold-coated, microporous nylon membranes as both the capture phase and the amperometric detector. The model system reported used a capture monoclonal antibody to hCG immobilised covalently to the gold surface via a self-assembled monolayer of thioctic acid. The separation-free assay was performed by incubating simultaneously the analyte (hCG) and a second, ALP-labelled, antibody with the antibody-coated electrode. The activity of specifically surface-bound ALP-Ab was resolved from excess ALP-Ab in the bulk solution by introducing the substrate (4-APP) from the reverse side of the porous membrane electrode. This allowed enzymic generation and oxidation of 4-AP formed in immediate proximity to the electrode surface. The assay performed effectively in whole blood.

Another recent approach to the development of amperometric electrochemical immunosensors capable of separation-free measurement has been to exploit the direct electrochemistry of enzymes immobilised at electrode surfaces using enzyme channelling or proximity effects (Maggio et al., 1980). McNeil's group has developed bioelectronic interfaces based on interfacial direct electron transfer to horseradish peroxidase at activated carbon electrodes, and has used these systems to detect H_2O_2 formed by enzyme-labelled species bund to electrode surfaces via specific recognition events (Ho et al., 1993; Ho et al., 1995; McNeil et al., 1995a; Wright et al., 1995). These generic systems have been applied to a number of analytes including a separation-free non-competitive assay for TSH. The TSH assay relied on H_2O_2 formation by an alkaline phosphatase label in the presence of the substrate 5-bromo-4-chloro-3-indolyl phosphate (Ho et al., 1995). Peroxide generated by the bound enzyme label was detected at the electrode surface by reduction with immobilized HRP followed by direct electron transfer from the activated carbon electrode to the active site of HRP. Enzyme channelling of H_2O_2 did not occur in the bulk solution.

Ivnitski and Rishpon (1996) have also developed an enzyme channelling method for the development of separation-free amperometric electrochemical immunosensors. In this system, capture antibodies and glucose oxidase were immobilised to graphite electrodes which had been coated with a polyethyleneimine film. The specific binding of HRP-labelled second antibodies, in a sandwich arrangement, was detected by the addition of glucose and iodide which caused enzyme channelling of H_2O_2, formed by the reaction of glucose with immobilised GOD, with concomitant oxidation of iodide to iodine. The overall system was monitored by electrochemical recycling of iodine formed by the HRP enzymic reaction, to iodide at a potential of −70 mV vs SCE. This system was applied to the one-step separation-free measurement of human luteinising hormone in serum.

Recent developments in electrochemical immunosensor technology have, in the main, concentrated on amperometric measurements. However, alternative electrochemical approaches should not be ignored. For instance McNeil et al. (1995b) have demonstrated a highly sensitive, electrochemical immunoassay method based on measurement of changes in electrode impedance at polymer-coated electrodes. Relatively little effort has been directed towards the exploitaton of electrode impedance measurements which have, potentially, distinct analytical and operational advantages including a dynamic range extending over four orders of magnitude a the lack of an absolute requirement for a reference electrode. In this approach (McNeil et al., 1995b) the impedance of electrodes coated with enteric polymer films was monitored as a function of the enzyme-catalysed dissolution of the film as a result of a local increase in pH due to the action of surface-bound urease-labelled antibodies on urea in the bulk solution. Enteric polymers are designed for coating of orally administered drugs for delivery to specific regions of the gastrointestinal tract. They are generally stable at acid pH and can be fabricated to dissolve at specific pH values above about 7. Once again this approach is based on a proximity effect, with the local chane in pH at the electrode surface causing dissolution of the molymer coating while not being affected by the bulk buffer capacity of blood samples. The impedance change involved in going from an electrode coated with an intact polymer film to a bare electrode is approximately four orders of magnitude.

Conclusion

Although many model systems have been described for electrochemical immunosensors with all of the necessary requirements for rapid, quantitative, one-step use where near-patient testing is highly desirable, if not essential, the world is still waiting with baited breath for real products which will supplant qualitative and semi-quantitative dry reagent immunoassay strip technologies. The next ten years should be very interesting.

References

Aizawa, M., Morioka, A., Susuki, S. and Nagamura, Y. (1979) Amperometric determination of hCG by membrane bound antibody. *Anal Biochem.* 94:22–28.

Athey, D. (1993) *Development of electrochemical enzyme immunoassays for thyrotropin.* PhD Thesis, University of Newcastle upon Tyne, UK.

Athey, D., Ball, M. and McNeil, C.J. (1993 a) Avidin-biotin based electrochemical immunoassay for thyrotropin. *Ann. Clin. Biochem.* 30:570–577.

Athey, D., McNeil, C.J., Bailey, W.R., Hager, H.J., Mullen, W.H. and Russel, L.J. (1993 b) Homogeneous amperometric immunoassay for theophylline in whole blood. *Biosens. Bioelectron.* 8:415–419.

Bartlett, P.N. and Whittaker, R.G. (1988) Strategies for the development of amperometric enzyme electrodes. *Biosensors* 3:359–379.

Bauer, C.G., Eremenko, A.V., Ehrentreich-Förster, E., Bier, F.F., Makower, A., Halsall, H.B., Heineman, W.R. and Scheller, F.W. (1996) Zeptomole detecting biosensor for alkaline phosphatase in an electrochemical immunoassay for 2,4-dichlorophenoxyacetic acid. *Anal. Chem.* 68:2453–2458.

Boyland, E. and Manson, D. (1957) The oxidation of aromatic amines Part IV oxidation by perphosphoric acids. *J. Chem. Soc.* IV:4689–4694.

Bresnahan, W.T. and Elving, P.J. (1981) Spectrophotometric investigation of products formed following the initial 1-electron electrochemical reduction of NAD^+. *Biochem. Biophys. Acta* 678:151–156.

Breyer, B. and Radcliffe, F.J. (1951) Polarographic investigation of the antigen antibody reaction. *Nature* 167:79.

Broyles, C.A. and Rechnitz, G.A. (1986) Drug antibody measurement by homogeneous enzyme immunoassay with amperometric detection. *Anal. Chem.* 58:1241–1245.

Cass, A.E.G., Davis, G., Francis, G.D., Hill, H.A.O., Aston, W.J., Higgins, I.J., Plotkin, E.V., Scott, L.D.L. and Turner, A.P.F. (1984) Ferrocene-mediated enzyme electrode for amperometric determination of glucose. *Anal. Chem.* 56:667–671.

deAlwis, W.U. and Wilson, G.S. (1985) Rapid sub-picomole electrochemical enzyme immunoassay for immunoglobulin G. *Anal. Chem.* 57:2754–2756.

diGleria, K., Hill, H.A.O., McNeil, C.J. and Green, M.J. (1986) Homogeneous ferrocene mediated amperometric immunoassay. *Anal. Chem.* 58:1203–1205.

Duan, C. and Meyerhoff, M.E. (1994) Separation-free sandwich enzyme immunoassays using microporous gold electrodes and self-assembled monolayer/immobilised capture antibodies. *Anal. Chem.* 66:1369–1377.

Eggers, H.M., Halsall, H.B. and Heineman, W.R. (1982) Enzyme immunoassay with flow amperometric detection of NADH. *Clin. Chem.* 28:1848–1851.

Elving, P.J., Bresnahan, W.T., Moirous, J. and Samec, Z. (1982) NAD/NADH as a model redox system; mechanisms, mediation, modification by the environment. *Bioelectrochem. Bioenerg.* 9:365–378.

Frew, J.E., Foulds, N.C., Wilshere, J.M., Forrow, N.J. and Green, M.J. (1989) Measurement of alkaline phosphatase activity by electrochemical detection of phosphate esters. *J. Electroanal. Chem.* 266:309–316.

Gil, E., Tang, H., Halsall, H., Heineman, W.R. and Misiego, A. (1990) Competitive heterogeneous enzyme immunoassay for theophylline by flow injection analysis with electrochemical detection of p-aminophenol. *Clin. Chem.* 36:662–665.

Glatz, J.F., Renneberg, R., McNeil, C.J. and Spener, F. (1995) Electrochemical and integrated-optical immunosensors for heart-type fatty acid binding protein – a new plasma marker for acute myocardial infarction. *In:* A. Pedotti and P. Rabischong (eds.) *Proceedings of 3rd European Conference on Engineering and Medicine,* Edizioni Pro Juventute Don Carlo Gnocchi, Roma, pp 179.

Heineman, W.R. and Halsall, H.B. (1985) Strategies for electrochemical immunoassay. *Anal. Biochem.* 57:1321–1331.

Hill, H.A.O. (1984) Assay techniques using specific binding agents. *European Patent Application* 84303090.9.

Ho, W.O., Athey, D., McNeil, C.J., Hager, H.J., Evans, G.P. and Mullen, W.H. (1993) Mediatorless horseadish peroxidase enzyme electrodes based on activated carbon: Potential application to specific binding assay. *J. Electroanal. Chem.* 351:185–197.

Ho, W.O., Athey, D. and McNeil, C.J. (1995) Amperometric detection of alkaline phosphatase at a horseradish peroxidase enzyme electrode based on activated carbon: Potential application to electrochemical immunoassay. *Biosens. Bioelectron.* 10:683–691.

Ivnitski, D. and Rishpon, J. (1996) A one-step, separation-free amperometric immunosensor. *Biosens. Bioelectron.* 11:409–417.

Jenkins, S.H., Halsall, H.B. and Heineman, W.R. (1991) Eclectic immunoassay. An electrochemical approach. *In*: A.P.F. Turner (ed.): *Advances in Biosensors*, Volume 1. JAI Press Ltd. London, pp 171–228.

Kulys, J., Razamas, V. and Malinauskas, A. (1980) Kinetic amperometric determination of hydrolase activity. *Anal. Chim. Acta.* 117:387–390.

Maggio, E.T., Wife, R.L. and Ullman, E.F. (1980) Reagents and method employing channelling. *United States Patent* 4233402.

Manning, P., Athey, D. and McNeil, C.J. (1994) Homogeneous amperometric immunoassay for digoxin. *Anal. Lett.* 27:2443–2453.

McNeil, C.J., Athey, D. and Ho, W.O. (1995a) Direct electron transfer bioelectronic interfaces: Application to clinical analysis. *Biosens. Bioelectron.* 10:75–83.

McNeil, C.J., Athey, D., Ball, M., Ho, W.O., Krause, S., Armstrong, R.D., Wright, J.D. and Rawson, K. (1995b) Electrochemical sensors based on impedance measurement of enzyme-catalysed polymer dissolution: Theory and applications. *Anal. Chem.* 67:3928–3935.

Meusel, M., Renneberg, R., Spener, F. and Schmitz, G. (1995) Development of a heterogeneous amperometric immunosensor for the determination of apolipoprotein-E in serum. *Biosens. Bioelectron.* 10:577–586.

Renneberg, R., Shlosser, W. and Scheller, F.W. (1983) Amperometric enzyme sensor-based enzyme immunoassay for factor VIII-related antigen. *Anal. Lett.* 16:1279–1289.

Robinson, G.A., Cole, V.M., Rattle, S.J. and Forrest, G.C. (1986) Bioelectrochemical immunoassay for human chorionic gonadotropin in serum using an electrode immobilised capture antibody. *Biosensors* 2:45–57.

Rosen, I. and Rishpon, J. (1989) Alkaline phosphatase as a label for a heterogeneous immunoelectrochemical sensor. *J. Electroanal. Chem.* 258:27–39.

Robinson, G.A., Cole, V.M., Rattle, S.J. and Forrest, G.C. (1986) Bioelectrochemical immunoassay for human chorionic gonadotrophin in serum using an electrode immobilised capture antibody. *Biosensors* 2:45–57.

Rubenstein, K.E., Schnieder, R.S. and Ullman, E.F. (1972) Homogeneous enzyme immunoassay. New immunochemical technique. *Biochem. Biophys. Res. Commun.* 47:846–851.

Siegmann-Thoss, C., Renneberg, R., Glatz, J.F.C. and Spener, F. (1996) Enzyme immunosensor for diagnosis of myocardial infarction. *Sensor. Actuator. B* 30:71–76.

Tang, H.T., Lunte, C.E., Halsall, H.B. and Heineman, W.R. (1988) p-Aminophenyl phosphate: an improved substrate for electrochemical immunoassay. *Anal. Chim. Acta* 214:187–195.

Thompson, R., Barone, III G., Halsall, B. and Heineman, W.R. (1991) Comparison of methods for following alkaline phosphatase catalysis; spectrophotometric versus amperometric detection. *Anal. Biochem.* 192:90–95.

van Nieuwenhoven, F.A., Kleine, A.H., Wodzig, K.W.H., Hermens, W.T., Kragten, H.A., Maessen, J.G., Punt, C.D., van Dieijen, M.P., van de Vusse G.F. and Glatz, J.F.C. (1995) Discrimination between myocardial and skeletal muscle injury by assessment of the plasma ratio of myoglobin over fatty acid-binding protein. *Circulation* 92:2848–2854.

Wright, J.D., Rawson, K.M., Ho, W.O., Athey, D. and McNeil, C.J. (1995) Specific binding assay for biotin based on enzyme channelling with direct electron transfer electrochemical detection using horseradish peroxidase. *Biosens. Bioelectron.* 10:495–500.

Frontiers in Biosensorics II
Practical Applications
ed. by. F.W. Scheller, F. Schubert and J. Fedrowitz

Biosensors based on flow-through systems

F. Spener, M. Meusel and C. Siegmann-Thoss

Institut für Chemo- und Biosensorik, D-48149 Münster, Germany

Summary. When combined with biosensors as the sensing element microdialysis and flow injection analysis (FIA) systems become sophisticated tools for handling analytical processes. In particular a FIA system offers a high degree of automation together with high reproducibility and small sample volumes, whereas the biosensor, allows selective and sensitive measurements of the various analytes. Here we describe first a miniaturised microdialysis flow-through system developed for glucose determination, then we focus on amperometric immunosensors and on microbial sensors. In the former, antibodies against low molecular weight environmental contaminants or against high molecular weight proteins are responsible for analyte detection, whereas the latter use immobilised microorganisms as the recognising element for monitoring water pollutants.

Introduction

Biosensors have become useful tools for solving analytical problems, especially when involved as sensing elements in microdialysis and flow injection analysis (FIA) systems. According to Ruzicka and Hansen (1988), a FIA system is based on the injection of a liquid sample into a moving, nonsegmented continuous carrier stream of a suitable liquid. The injected sample forms a zone, which is then transported toward a detector that continuously records the absorbance, electrode potential, or other physical parameters as it continuously changes due to the passage of the sample material through the flow cell. A detailed discussion of FIA principles and theory is beyond the scope of this article and for a more comprehensive treatment of the basics of FIA, the second edition of "Flow Injection Analysis" by Ruzicka and Hansen (1988) and other monographs on this topic (Fang, 1993) may be consulted.

The major advantages of these systems are a short contact time between the analytes and the biosensors, a high sample throughput and small sample volume, often without any sample pre-treatment. Moreover, the reproducibility is extremely high and calibrations are easy to perform. Due to these reasons FIA is rapidly gaining the attention of both researchers and engineers interested in the application of biosensors for bioprocess monitoring.

Many applications of biosensors integrated into flow-through systems for process monitoring have been described. Busch et al. (1993) used optical sensing principles based on an urea optode and a glucose luminescence sensor for a simultaneous analyte determination. Optodes as well as an enzyme thermistor were applied for monitoring penicillin V concentration during cultivation of *Penicillium chrysogenum* and for the determination of various mouse IgGs, respectively (Scheper et al., 1993). Besides optical transducers, amperometric devices are commonly used in process control. Amperometric biosensors for glucose, glutamine and glutamate were incorporated into a three-cell parallel FIA system used to monitor the analytes on-line during mammalian cell perfusion cultures (White et al., 1995). Céspedes et al. (1995) used a glucose biosensor in a FIA system for fermentation monitoring. In this case a biocomposite material made of graphite-epoxy-Au-Pd-glucose oxidase was integrated in a flow injection system and used for amperometric glucose determination.

Apart from process control, flow-through systems are extensively used in biosensor research and development. This may be due to the versatility of flow operations, which are easy to automate, easy to control in space and time, and the underlying fluidics can be miniaturised. Flows can be mixed, stopped, restarted, reversed, split, recombined and sampled (Ruzicka and Hansen, 1988). Finally, flow operations allow most detectors and sensors to be used in a more reproducible manner than in batch operations by hand. As in the systems for process control, among the transducers optical and electrochemical principles dominate. Hansen et al. (1993) reported an enzyme FIA using three coimmobilised enzymes for the fluorometric detection of trace levels of ATP. The amplification scheme for ATP is based on the enzymes pyruvate kinase, hexokinase and glucose-6-phosphate dehydrogenase, resulting in the generation of NADH which is measured fluorometrically. A bioluminescence-based fiber optic sensor for flow injection analysis was described by Blum et al. (1993). Firefly luciferase, specific for ATP, and the bacterial oxidoreductase/luciferase system, specific for NADH, have been immobilised on preactivated membranes. By co-immobilising both bioluminescence systems, a multi-function biosensor was designed. Exploiting the chemoluminescence of luminol, Preuschoff et al. (1993) developed a hydrogen peroxide sensor for use in flow analysers.

Besides optical detectors, amperometric detectors are most commonly used in enzyme-FIA systems. Amperometric enzyme sensors for the determination of lactate, glutamate and glutamine were described by White et al. (1994). An enzyme-FIA system for galactose determination was proposed by Manowitz et al. (1995), using galactose oxidase immobilised on a platinised carbon electrode. To prevent interferences from human plasma the electrode was modified with a composite polymer.

At present only few commercial bioanalytical FIA-systems are available, e.g. enzyme-FIA systems from Anasyscon (Hannover, Germany) or Ismatec (Glattbrugg, Switzerland). These systems consist of the FIA hardware and exchangeable enzyme cartridges. In addition, during the last years Biochemical Oxygen Demand (BOD)-analysers based on microorganisms as sensing elements in a FIA setup have been commercialised, showing clearly the demand for automated sensor systems (BOD-Module, Prüfgeräte-Werk Medingen GmbH, Dresden, Germany; ARAS Sensor BOD-System, Dr. Bruno Lange GmbH, Düsseldorf, Germany; BODypoint, Aucoteam GmbH, Berlin, Germany, BOD-2000/BOD-2200, Central Kagaku Corp., Tokyo, Japan).

The present paper deals with recent developments of biosensors based on flow-through systems. Each of the developments is tailored for a certain application using sensor technology from our institute and generic approaches whenever possible. This report focuses on glucose determination in a miniaturised microdialysis system and on FIA-systems making use of immunosensors and microbial sensors as detection units.

Microdialysis enzmye system for glucose monitoring

In human medicine a continuous monitoring of different parameters is often required. For instance the monitoring of blood glucose as an aid for the treatment of diabetes could improve the process of dose optimisation in the beginning of the medical treatment. In our institute a continuously working microdialysis flow-through system based on the enzyme glucose oxidase as recognition element has been developed for glucose monitoring (Knoll, 1995 a).

Until now the enzyme membrane in micro-enzyme sensors has been deposited on the top of the transducer. Problems are often caused by membrane adhesion and the mechanical stability of the membrane. To solve these problems a new concept for membrane deposition, the so-called 'containment' technology (Knoll, 1995 b), has been developed where the enzyme membrane is deposited on the chip in pyramidal containments produced on silicon by anisotropic etching. These containments show opening sides towards the analyte solution of 120 μm. The enzyme membrane, a photocrosslinkable polyvinylalcohol with glucose oxidase, was deposited inside this containment.

Figure 1 shows schematically the miniatuarised flow-through system (Cammann et al., 1994). A carrier solution is pumped continuously through a microdialysis needle by a syringe driven by gas pressure. To achieve a constant flow rate which is independent of the actual gas volume a liquid/gas system has been used. The tip of the microdialysis needle is in contact with the subcutaneous tissue. It consists of a dialysis membrane preventing macromolecules and other macrocomponents to enter the carrier solution whereas low molecular weight components are allowed to diffuse through the membrane into the carrier solution. The flow rate of the system is adjusted by two capillary channels which are integrated in the flow-through sensor chip, the heart of the system. It consists of the enzyme-containing sensor elements and the two capillary channels connected by a third channel. The total chip dimensions are 11 mm × 6 mm.

a)

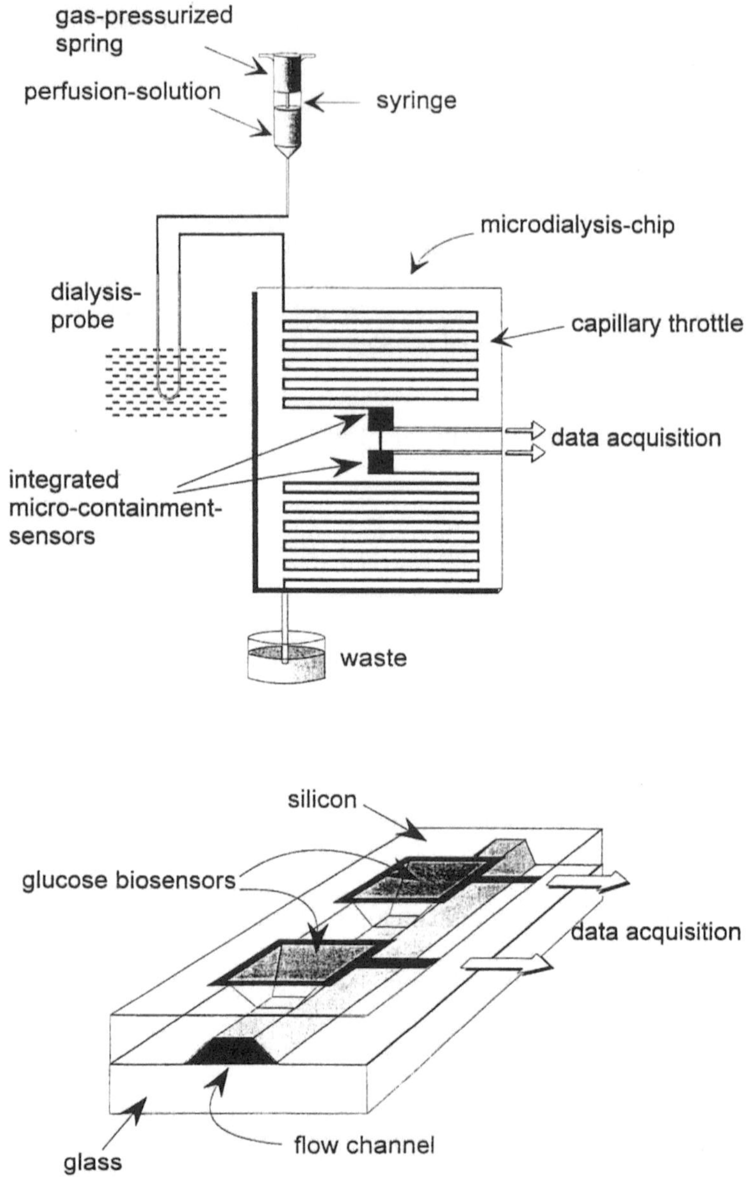

b)

Figure 1. Schematic view of the microdialysis flow-through glucose monitoring system; a) general view with the pumping unit, the microdialysis sampling, the flow-through sensor chip and electronics; b) close-up view of the sensitive part.

The amperometric measurement bases on the enzymatic oxidation of glucose with the enzyme glucose oxidase (GOD). The hydrogen peroxide generated in this reaction can be electrochemically oxidised at a platinum anode, which is polarised at + 600 mV vs Ag/AgCl. The resulting current is measured with a miniaturised amplifier electronics with a LCD display which has been developed to get a hand-held monitoring system with overall dimensions of a cigarette pack.

Although problems currently arise from interfering substances it could be demonstrated that the sensor exhibits a linear range wide enough to cover the complete glucose concentration range of diabetic people. The device has been teste *in vitro* in human serum and glucose standard solutions. It works over a period of at least 5 days without loss of sensitivity. *On patient* measurements will be carried out in due course.

Due to its miniaturisation the flow-through system opens the chance for continuous *in vivo* monitoring of blood glucose. This is a first step towards the development of a miniaturised artificial pancreas consisting of the glucose monitoring system coupled with an appropriate insulin delivery system. In contrast to implantable needle-type sensors this microsystem including a flow-through cell avoids many problems related to biocompatibility (Steinkuhl et al., 1994). Beside these advantages containment sensors in general are very robust, show a good response time and a long life time, and their fabrication process is entirely compatible with mass-production technologies since it is a full wafer process (Steinkuhl et al., 1996).

Immunosensor systems

The use of antibodies as biological components offers the possibility for highly sensitive and selective measurements, especially in complex matrices. The specificity of the sensor systems is principally governed by the properties of the antibody molecule. The sensitivity of the device, however, is dependent on both, the biological component and the transducer. By selecting the antibody molecule the sensor system can be adapted and optimised for specific purposes. For single-compound determinations an antibody without crossreactivities towards structurally similar substances will be most appropriate, whereas for sum-value estimations an antibody with a wider spectrum of target molecules will be the antibody of choice.

In analytical immunochemistry the traditional ELISA has earned its value as an accurate, non-radioactive method with impressive detection limits. Yet it is a time-consuming, labour-intensive, off-line method that is commonly regarded as being too slow for process monitoring and control. By adapting the ELISA-principle to FIA systems it has become possible to overcome many drawbacks of this technology (Middendorf et al., 1993; Nilsson et al., 1993). With this is mind we developed an immunosensor system for the determination of both haptens and high molecular weight antigens.

Determination of herbicides in drinking water

The herbicide 2,4-D (2,4-dichlorophenoxyacetic acid) was chosen as a prototpye analyte, because phenoxyacetic acids have been, and in some parts of the world, still are being used to a great extent as chemical agents in agricultural weed control.

Naturally the herbicides applied as well as their catabolites are prone to contaminate not only ground and surface waters, but drinking water as well. As a consequence EU guidelines for drinking water quality allow 0.1 μg/l as the highest concentration permitted for a specific herbicide. Conventional methods for the determination of such analytes are based on chromatographic methods (GC and HPLC) requiring complicated and time-consuming sample pre-treatments. HPLC-analysis, for example, needs a sample volume of 2 l followed by a solid-phase extraction with organic solvents, a time-consuming procedure of several hours. Recently, a disposable amperometric immunosensor for 2,4-D determination was described (Kaláb and Skládal, 1995). This sensor is based on a screen-printed electrode system as the amperometric transducer and monoclonal antibodies against 2,4-D as the biospecific part. In contrast to these analytical methods an automated flow-through system meets the requirements of continuous monitoring of drinking water.

The computer-controlled FIA arrangement developed here includes an immunoreactor, a potentiostat, a peristaltic pump, a selector and a sampler (Fig. 2). The principle of the assay format is depicted in Figure 3. During a preincubation step 2,4-D in the sample binds to an alkaline phosphatase (AP)-labelled monoclonal anti-2,4-D antibody. This 'cocktail' format was successfully tested in an ELISA (Wilmer et al., 1996a) as well as in a Sequential Injection Analysis (SIA)-arrangement (Wilmer et al., 1996b). Since the antibody is added as a soluble reagent, its concentration can be well controlled and its natural affinity is preserved. When this mixture is applied to the immunoreactor only the antibodies that have not previously bound 2,4-D can bind to the hapten which is immobilised covalently on the surface of the reactor. All unbound species are washed away by the carrier stream buffer. Enzyme activity bound to the reactor is detected by applying substrate (p-aminophenyl phosphate) solution to the reactor and monitoring the product downstream with an appropriate flow-through electrochemical detector. After detecting bound activity, the reactor is regenerated by running a low pH buffer through the column for a brief period. As the amount of labelled antibodies that bind to the reactor decreases with increasing 2,4-D concentrations the signal generated in the sensor is inversely proportional to the concentration of 2,4-D in the sample. The time required for one measurement is 12 min.

For the immobilisation of 2,4-D on the surface of the immunoreactor, 2,4-D was coupled to bovine serum albumin (BSA) via the carboxylate group. In first experiments the SiO_2 surface of a capillary immunoreactor

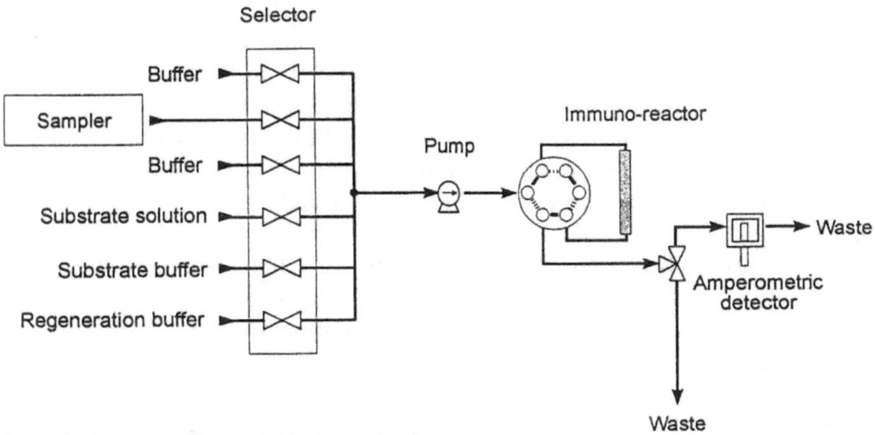

Figure 2. FIA setup for pesticide determination.

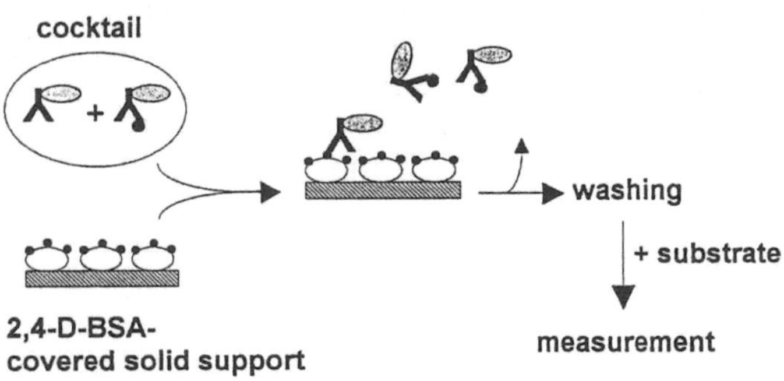

Figure 3. Immunosensing principle, (•, analyte; antibody-enzyme-conjugate).

was activated with tosyl chloride. Coupling of 2,4-D-BSA-conjugates was then carried out with the help of a carbodiimide. The operational stability of this surface was limited to about 20 regeneration cycles. The stability of the column could be significantly increased, however, by using poly-L-lysine instead of BSA of 2,4-D coupling.

The electrochemical detection of alkaline phosphatase based on the hydrolysis of phenylphosphate to phenol followed by phenoloxidation at + 670 mV vs Ag/AgCl was first described by Heineman and co-workers who developed heterogeneous immunoassays based on this principle (Wehmeyer et al., 1985). Later on, p-aminophenyl phosphate was proposed as improved substrate (Tang et al., 1988) as it can be oxidised at considerably lower potentials, around + 100 mV, thus avoiding interferences common in many real matrices (Meyerhoff et al., 1995). We have applied this generic detection principle successfully to matrices like water, serum or urine.

p-Aminophenyl phosphate is not commercially available and was therefore synthesised according to DeRiemer and Meares (1981). The K_m-value of p-aminophenyl phosphate in the alkaline phosphatase reaction is 267 μM, for use in the FIA-system this substrate was dissolved in carbonate buffer pH 9.6 and added in 7-fold excess to assure saturation kinetics. After enzymatic conversion, p-aminophenol was oxidised at a graphite electrode of 3 mm diameter at a potential of +150 mV vs Ag/AgCl.

An important factor in the development of a FIA-arrangement is the optimisation of the hapten-antibody interaction in the immunoreactor. In the titration assay the choice of a suitable concentration of the labelled antibody and the flow-through time of this conjugate is most important. An increase of both parameters results in a linear increase of the sensor signal before reaching a plateau. In both cases the parameters have to be chosen such that the signal generated is surely within the linear range. Otherwise, small changes in the analyte concentration would not lead to a significant sensor signal. With the optimised system a calibration curve for 2,4-D in the concentration range of 0.1 to 50 μg/l was obtained (Fig. 4). The detection limit of 0.1 μg/l matches with EU standards, however, a lower detection limit would be appropriate. Experiments with 2,4-D determination in spiked water samples indicate that such a low detection limit can be

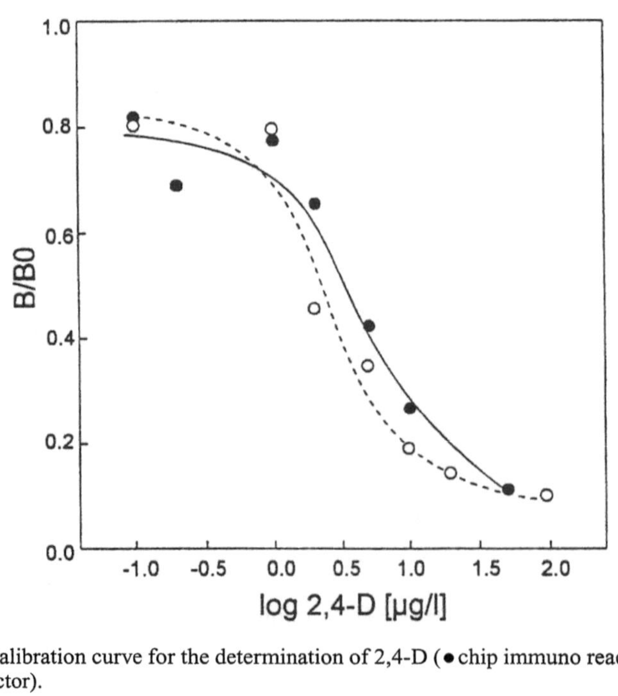

Figure 4. Calibration curve for the determination of 2,4-D (● chip immuno reactor; ○ capillary immunoreactor).

reached by optimising the antibody-antigen reaction and the electrochemical detection without using laborious and costlier amplification systems.

With our aim to provide low-cost sensor systems by miniaturisation we introduced silicium chip technology to replace the capillary immunoreactor. As the surface of the silicium chips also consists of SiO_2 similar chemistry for coupling of the hapten could be applied. Calibration curves obtained with this miniaturised devices (Fig. 4) are comparable to those of large and somewhat delicate glass capillaries. Sensor miniaturisation and integration of multiple transducers require the employment of microfabrication technology (Xie et al., 1995). For many future applications and especially for the concept of 'on-chip biosensors' these miniaturised devices are indispensable. In future developments even microfluidics, filtres or micro-valves will be integrated 'on-chip'. Current developments at ICB focus on such miniaturisations of electrodes and electrode arrays for the application in immuno and enzyme flow-through systems.

Polycyclic aromatic hydrocarbons

One of the major advantages of immunosensors is the versatility of these devices. As the specificity of the sensor is almost entirely governed by the biological component, the system can be easily adapted to another analyte by changing the biological recognition element. In this case the flow through arrangement and the titration principle as described before was used for the development of an immunosensor for polycyclic aromatic hydrocarbons (PAH) in methanolic soil extracts. Many of the PAH are hazardous substances due to their potential deleterious effects on human health and ecosystems. In contrast to single-compound determinations the PAH-sensor was optimised for the estimation of a sum-value, using polyclonal antibodies raised against phenanthrene which crossreacted with other PAH. The estimation of an overall concentration below or above a certain threshold is an ideal parameter for screening purposes, especially in the environmental field. By selecting the antibody best suited, or by using an antibody cocktail, automated immunosensors based on FIA-systems can be developed as a practical tool for the rapid, easy and cheap analysis of a large number of samples.

The most common method for PAH determination is HPLC-analysis with fluorescence/UV-detection after soxhlet or supercritical fluid extraction (SFE). Driven by the need for faster and more costeffective methods for environmental monitoring, a variety of environmental field analytical methods are commercially available or currently being developed. Especially field test kits are relatively mature, resulting in a number of commercial products. However, for several applications, especially for field monitoring, a demand for repetitive or continuous monitoring of samples exists (Rogers et al., 1995).

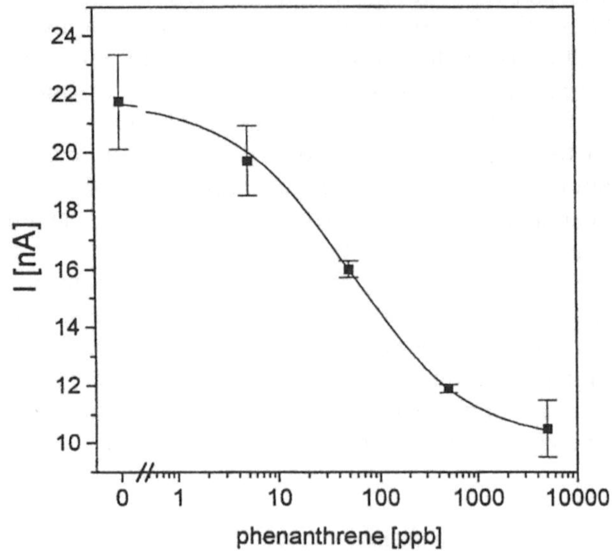

Figure 5. Calibraton curve for the determination of phenanthrene (capillary immunoreactor).

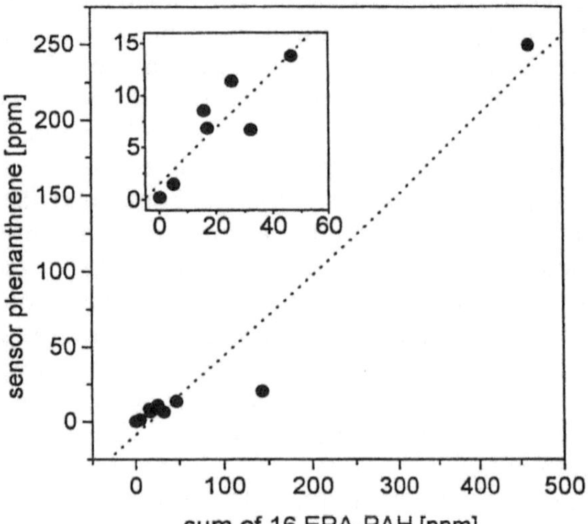

Figure 6. Determination of PAH extracted from soil samples. Comparison of sensor measurements (FIA-system with immunoreactor) and the routine HPLC-method.

Following the principle of the immuno-FIA-system described above on the surface of the immunoreactor phenanthrene was covalently immobilised via BSA or poly-L-lysine. As the polyclonal antibodies raised against phenanthrene were not enzyme-labelled at this stage of the investigations a secondary AP-labelled antibody was used for the detection of the hapten-bo und antibodies. The calibration curve for the determination of phenanthrene shown in Figure 5 indicates that this hazardous substance can be determined within a linear range of 5 to 500 ppb.

As the sensor was tailored in particular for the PAH determination in soil, the choice of an appropriate procedure for sample extraction was very important. In this case, several samples of contaminated soil were extracted with methanol for 1 min. Due to the extremely low detection limit of the sensor calibrated with phenanthrene the extracts could be applied to the sensor after diluting with phosphate buffer only, without any further pretreatment. To estimate the precision of the immunochemical method HPLC-results were compared to those obtained with the immunosensor (Fig. 6). The concentration of 16 PAH in 7 soil samples was determined according to the US Environmental Protection Agency (EPA)-method 8310. Figure 6 shows a good correlation between the sum-value of the immunosensor and the sum of the 16 EPA-PAH, when the latter empirically was multiplied by 2. Prerequisites for such an operation were the cross-reactivity of the polyclonal anti-phenanthrene-antibody with the other PAH and a constant concentration ratio of main phenanthrene to the other PAH in the soil samples. The former was verified by ELISA, the latter confirmed in preceding HPLC measurements of the various soil samples.

In further investigations the concentration of PAH in the urine of several people who were exposed to high levels of PAH was determined with the immunosensor, however, in this case no correlation between the GC/MS-data of single compounds and the sum-value of the sensor could be observed. Due to metabolisation and modification of the PAH, e.g. glycosylation, the composition of these substances in urine is completely different than that in soil. The concentration of unmodified phenanthrene was too low to serve as a meaningful leader analyte.

Determination of monoclonal mouse-IgG

The versatility of the system was further demonstrated by using the sensor not only for the determination of haptens like 2,4-D or PAH but also for proteins such as monoclonal antibodies. The monoclonal antibody production in hybridoma cultures is of special importance. Here, on-line monitoring can be beneficial (Mulchandani and Bassi, 1995). Due to the addition of fetal calf serum (FCS) during hybridoma culture the amount of specific monoclonal antibodies is often significantly lower than the total antibody concentration. FCS, which is added in a concentration of 5 to 10%, contains a high amount of bovine antibodies and therefore leads to a misrepresentation of the concentration of specific antibodies. To avoid this problem the monoclonal antibody-specific antigen can be immobilised on the surface of the immunoreactor. In this case only specific antibodies in the sample will bind to the reactor.

For the determination of monoclonal anti-2,4-D antibodies the hapten was immobilised via BSA on the surface of the capillary immunoreactor as described before. First, the antibodies in the sample bind to the hapten in the reactor. Afterwards, the antibodies were detected using an AP-labelled secondary antibody recognising the heavy chain of mouse-IgG. The system is very easy to use and needs no complex optimisation procedures as in this assay format both the hapten on the surface and the secondary antibody are used in excess. It is another advantage of this format that the signal generated is directly proportional to the analyte concentration. The calibration curve once taken is valid for the operational life-time of the immuno-reactor, the binding capacity of the reactor needs to be redetermined, however, to check the linear range. The only parameter that needs careful optimisation is the pH of the sample, because the pH-value has a significant influence on the binding of the antibodies. The detection limit for monoclonal anti-2,4-D IgGs was determined to approximately 10 ng/ml (data not shown). Time required for one measuring cycle was 10 min.

Microbial sensors

Microbial sensors using immobilised microorganisms as recognition elements show some advantages towards biosensors using proteins such as antibodies or enzymes. They are less sensitive to inhibition and more tolerant of suboptimal pH and temperature and often have a longer lifetime. In addition, these sensors are cheaper because they employ microorganisms instead of proteins which have to be isolated and purified (Karube and Suzuki, 1990). Furthermore the whole cell may perform multistep transformations and is able to recognise a group of substances, however, at the cost of a diminished selectivity. Due to the huge wealth of microorganisms with their wide spectrum of metabolic types microbial sensors are an inexhaustible reservoir for many applications (Riedel, 1994).

Microbial sensing of naphthalene

In the framework of the discussion about species analysis microbial sensors may be the best solution for the determination of the biologically available amount of the analyte especially in aqueous samples. For the determination of overall amounts of analyte, however, sensor systems based on sample pretreatment or extraction are better suited.

With this in mind an amperometric microbial flow-through biosensor was developed for the determination of biologically available PAH in aqueous

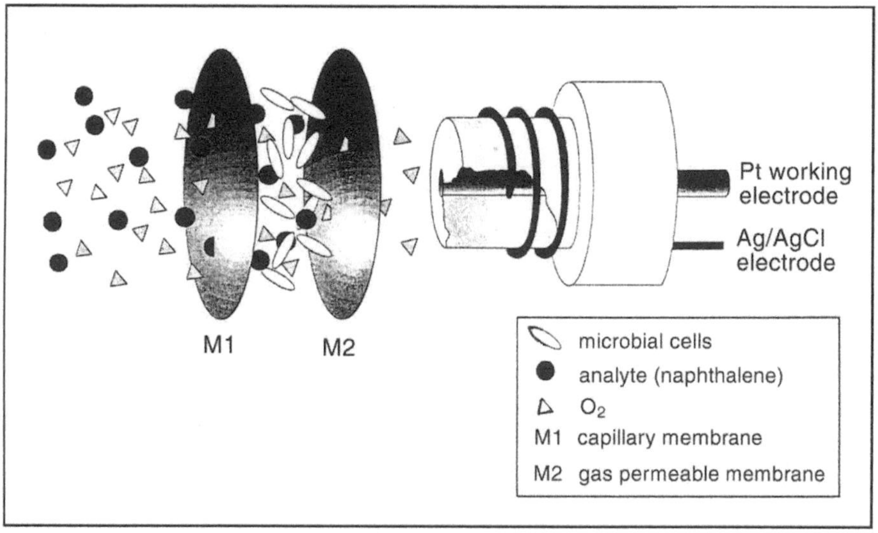

Figure 7. Schematic setup of the amperometric microbial biosensor.

samples using either immobilised *Sphingomonas* sp. B1 or *Pseudomonas fluorescens* WW4 cells that are able to use PAH as the sole carbon source while consuming oxygen (König et al., 1996). Both strains showed an identical performance in the determination of naphthalene in aqueous solutions.

An oxygen electrode was first covered with a gas permeable membrane onto which the immobilised microorganisms were adhering and then with a capillary membrane. The microorganisms were immobilised in a polyurethane-based hydrogel (Vorlop et al., 1992). This sensor was installed in a flow-through system (BOD-Module, PGW Medingen GmbH, Dresden, Germany) with an alternating flow of buffer and analyte solution. Figure 7 shows the schematic setup of the biosensor. During the measurement the current decreases for 2 min while the analyte solution is passed through the system. The assimilation of naphthalene by the immobilised microorganisms reduces the oxygen flow to the electrode. After stopping the contact between sensor and analyte the current reincreases.

The measurements in the flow-through system resulted in sensitivities of 0.5 nA/(mg/l) naphthalene, a background noise of ±0.01 nA and a linear range from 0.06 to 2.0 mg/l. Actual results show that an increase of sensitivity to values up to 1.2 nA/(mg/ml) and a decrease of detection limit down to 0.03 mg/l is possible. The reproducibility was within ±5% of the mean value in a series of ten samples when the test solution contained 1.5 mg/l naphthalene (Fig. 8). The lifetime of the sensor was 20 days with a sensitivity decrease of 50% of the initial value. In comparision with these results, measurements in a stirred cell (batch method) showed sensitivities between 3 and 4 nA/(mg/l), a background noise of ±0.01 nA and a linear range from 0.01 to 3.0 mg/l.

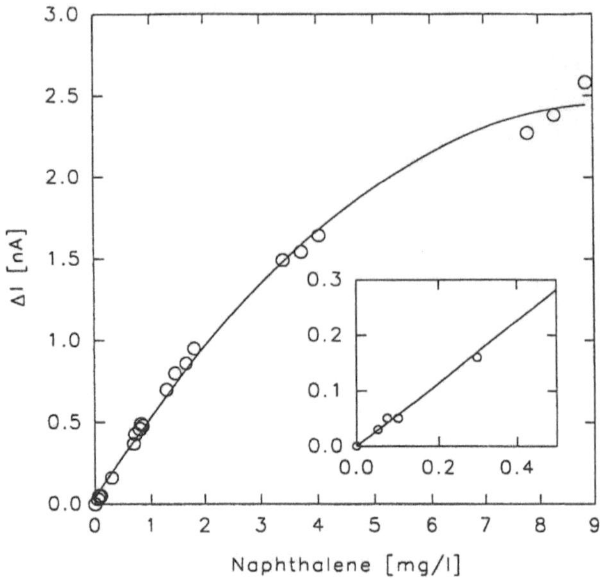

Figure 8. Calibration curve for the determination of naphthalene using the microbial sensor with *Sphingomonas* sp.

Both sensors reached lower detection limits than most microbial sensors for xenobiotics described so far (Riedel et al., 1991, 1993). In comparison to the batch system the flow-through system shows a reduced sensitivity and detection limit due to the fact that the sensor is supplied with oxygen by injection of air into the flow of buffer and analyte solution. The subsequent aerosol formation leads to a 75% loss of volatile naphthalene before the analyte reaches the sensing unit. Nevertheless the impressive detection limits are due to the high naphthalene degradative capacities of the microorganisms as well as to the low background noise, the latter can be attributed to the homogeneous diffusion properties of the polyurethane based hydrogel matrix which was used for the first time in a microbial biosensor (König et al., 1996).

The selectivity of the sensor was determined by taking calibration curves for a variety of substrates and calculating the sensitivities within the linear range in relation to that for naphthalene. These results show that the selectivity of the *Sphingomonas* sp. containing sensor is generally better than the selectivity of the sensor with *P. fluorescens* (data not shown).

With this sensor a simple and rapid method for the determination of naphthalene in aqueous solutions was developed. The linear range of this sensor corresponds with concentrations found in aqueous extracts of contaminated soil (up to 1 mg/l) and matches PAH threshold values for the remediation of ground water stated in the *List of the Netherlands* (0.04 mg/l, VROM/NL, 1988).

Due to the high and automated sample throughput, the flow-through system is the more promising tool for on-line measurements and screening of aqueous samples. On account of the short contact time of the sample with the sensor a growth of the microorganisms and an induction of additional enzymes is less likely and therefore performance and selectivity of the sensor are maintained.

Sensor for rapid estimation of the biochemical oxygen demand (BOD) using heavy metal resistant microorganisms

The biochemical oxygen demand (BOD) is a widely used parameter in environmental water control. It indicates the cotent of biodegradable organic compounds in waste water. To measure the conventional BOD_5 takes five days and is certainly not practicable for rapid and actual monitoring of waste water. The BOD can be estimated more rapidly by using the microbial sensor format described before. The general problem that heavy metal ions can be present in waste water and cause inhibitory effects on the microorganisms in the sensor can be avoided by using heavy metal resistent strains.

We developed a sensor using *Alcaligenes eutrophus* KT02 which carries plasmids encoding resistance to 40 mM $NiCl_2$, 20 mM $CoCl_2$, 10 mM $ZnCl_2$ and 1 mM $CdCl_2$ (Slama et al., 1995). The cells were immobilised via entrapment in the polyurethane hydrogel. All measurements were carried out with the microbial sensor integrated in the flow-through system and showed that the estimation of BOD even in presence of 4 mM nickel, copper and zinc is possible. Cadmium, which is normally not present in the environment in high concentrations, is tolerated by this strain in a concentration of 4 mM for a period of 4 h. A typical current response of this sensor towards various amounts of glycerol during a series of measurements is shown in Figure 9.

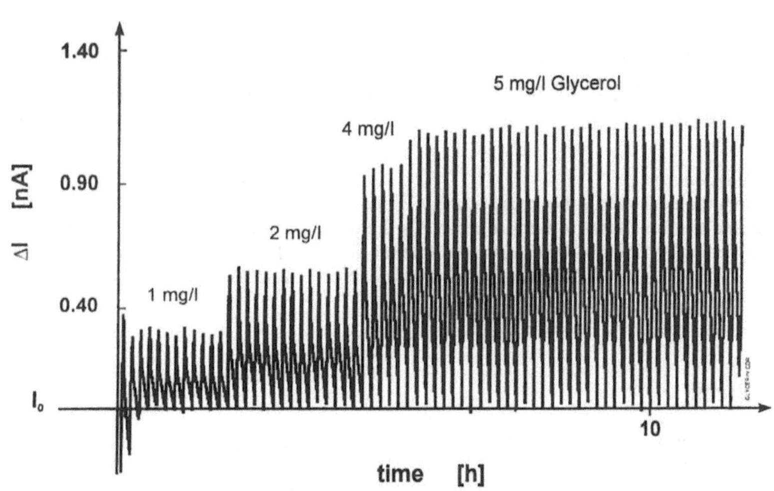

Figure 9. Response of a microbial sensor with *Alcaligenes eutrophus* KT02 carrying a plasmid encoding heavy metal resistence to glycerol in the presence of 4 mM Cd^{2+}

The sensor shows a lifetime of more than 30 days allowing a sample throughput of 7.5 per hour.

Conclusions

The specific features of biosensors include advantages over conventional analytical techniques such as (i) the specificity of the sensor that is principally governed by the biological component, (ii) fast measurements through the use of high sample throughput in automated analysers and (iii) continuous measurements (Mulchandini and Bassi, 1995). These advantages are especially exploited in flow-through systems and therefore the number of applications and developments is not surprising. FIA systems including biosensors as detecting elements are used in sensor research and development as well as in process control.

As demonstrated, amperometric biosensors associated with flow injection analysis have been successfully developed. On the basis of these results automatic or semi-automatic devices will enable the continuous monitoring of a specific compound, which would be particularly useful in industrial processes and environmental control. In particular for the latter the development of continuous and *in situ* monitoring techniques remains a key concern and, in addition to single-compound-measurements, the estimation of sum-values will provide excellent data for screening purposes.

One of the major immanent advantages of flow-through systems is the variety of uses which was exemplified here by (i) a miniaturised microdialysis enzyme flow-through system for continuous glucose monitoring, (ii) a microbial sensor integrated in a flow-through system, and (iii) immuno-FIA-systems for the determination of both haptens and proteins. In contrast to routine analysis methods these systems offer a high degree of versatility, simple reagent requirements and almost require no sample pre-treatment.

Acknowledgements
The authors thank the Fraunhofer Institut für Festkörpertechnologie (Munich) for supplying the miniaturised silicon-made immunoreactor. For the work carried out at the ICB financial support from the Federal Ministry of Education, Science, Research and Technology (BMBF) and the Ministry of Science and Research of the State Northrhine Westfalia (MWF) is gratefully acknowledged.

References

Blum, L.J., Gautier, S.M. and Coulet, P.R. (1993) Design of bioluminescence-based fiber optic sensors for flow-injection analysis. *J. Biotechnol.* 31:357–368.

Busch, M., Höbel, W. and Polster, J. (1993) Software FIACRE: Bioprocess monitoring on the basis of flow injection analysis using simultaneously a urea optode and a glucose luminescence sensor. *J. Biotechnol.* 31:327–343.

Camann, K., Hinkers, H. and Knoll, M. (1994) Microstructures and microsystems in instrumental analysis. *Analusis* 22:M19–M21.
Céspedes, F., Valero, F., Martinez-Fàbregas, E., Bartroli, J. and Alegret, S. (1995) Fermentation monitoring using a glucose biosensor based on an electrocatalytically bulk-modified epoxy-graphite biocomposite integrated in a flow system. *Analyst* 120:2255–2258.
DeRiemer, L.H. and Meares, C.F. (1981) Synthesis of mono- and dinucleotide photoaffinity probes of ribonucleic acid polymerase. *Biochemistry* 20:1606–1612.
Fang, Z. (1993) *Flow injection separation and preconcentration*. VCH, Weinheim, Germany.
Hansen, E.H., Gundstrup, M. and Mikkelsen, H.S. (1993) Determination of minute amounts of ATP by flow injection analysis using enzyme amplification reactions and fluorescence detection. *J. Biotechnol.* 31:369–380.
Kaláb, T. and Skládal, P. (1995) A disposable amperometric immunosensor for 2,4-dichlorophenoxyacetic acid. *Anal. Chim. Acta* 304:361–368.
Karube, I. and Suzuki, M. (1990) Microbial biosensors. *In*: A.E.G. Cass (ed.): *Biosensors – A Practical Approach*. Oxford University Press, Oxford, U.K., pp 155–170.
König, A., Zaborosch, C. Muscat, A., Vorlop, K.-D. and Spener, F. (1996) Microbial sensors for naphthalene using *Sphingomonas* sp. B1 or *Pseudomonas fluorescens* WW4. *Appl. Microbiol. Biotechnol.* 45:844–850.
Knoll, M. (1995a) Miniaturisiertes Durchflußanalysesystem. *German Patent* P4410224.0-52.
Knoll, M. (1995b). Verfahren zur Herstellung von miniaturisierten Chemo- und Biosensorelementen mit ionenselektiver Membran sowie von Trägern für diese Elemente. *German Patent* DE 4115414A1.
Manowitz, P., Stoecker, P.W. and Yacynych, A.M. (1995) Galactose biosensors using composite polymers to prevent interferences. *Biosens. Bioelectron.* 10:359–370.
Meyerhoff, M.E., Duan, C. and Meusel, M. (1995) A novel non-separation sandwich type electrochemical enzyme immunoassay system for detecting marker proteins in undiluted blood. *Clin. Chem.* 41:1378–1384.
Middendorf, C., Schulze, B., Freitag, R., Scheper, T., Howaldt, M. and Hoffmann, H. (1993) Online immunoanalysis for bioprocess control. *J. Biotechnol.* 31:395–403.
Mulchandini, A. and Bassi, A.S. (1995) Principles and applications of biosensors for bioprocess monitoring and control. *Crit. Rev. Biotechnol.* 15:105–124.
Nilsson, M., Mattiasson, G. and Mattiasson, B. (1993) Automated immunochemical binding assay (flow-ELISA) based on repeated use of an antibody column placed in a flow-injection system. *J. Biotechnol.* 31:381–394.
Preuschoff, F., Spohn, U., Blankenstein, G., Mohr, K.-H. and Kula, M.-R. (1993) Chemiluminometric hydrogen peroxide sensor for flow injection analysis. *Fresenius J. Anal. Chem.* 246:924–929.
Riedel, K. (1994) Microbial sensors and their applications in environment. *Exp. Technique Phys.* 40: 63–76.
Riedel, K., Naumov, A.V., Boronin, A.M., Golovleva, L.A., Stein, H.J. and Scheller, F. (1991) Microbial sensors for determination of aromatics and their chloroderivatives I. Determination of 3-chlorobenzoate using a *Pseudomonas*-containing biosensor. *Appl. Microbiol. Biotechnol.* 35:559–562.
Riedel, K., Hensel, J., Rothe, S., Neumann, B. and Scheller, F. (1993) Microbial sensors for determination of aromatics and their chloroderivatives II. Determination of chlorinated phenols using a *Rhodococcus*-containing biosensor. *Appl. Microbiol. Biotechnol.* 38: 556–559.
Rogers, K.R. (1995) Biosensors for environmental applications. *Biosens. Bioelectron.* 10: 533–541.
Ruzicka, J. and Hansen, E.H. (1988) *Flow injection analysis,* 2nd edn. John Wiley & Sons, New York, N.Y.
Scheper, T., Brandes, W., Maschke, H., Plötz, F. and Müller, C. (1993) Two FIA-based biosensor systems studies for bioprocess control. *J. Biotechnol.* 31:345–356.
Slama, M., Zaborosch, C. and Spener, F. (1995) Microbial sensor for rapid estimation of the biochemical oxygen demand (BOD) in presence of heavy metal ions. *In*: R.D. Wilken, A. Knöchel and U. Förstner (eds.): *Int. Conference Heavy Metals in the Environment*, Vol. 2. CEP Consultants Ltd., Edinburgh, U.K., pp 171–174.
Steinkuhl, R., Hinkers, H., Dumschat, C., Knoll, M. and Camman, K. (1994) Glucose sensor in containment technology. *Horm. Metab. Res.* 26:531–533.

Steinkuhl, R., Dumschat, C., Sundermeier, C., Renneberg, R., Cammann, K. and Knoll, M. (1996) Micromachined glucose sensor. *Biosens. Bioelectron.* 11:187–190.

Tang, H.T., Lunte, C.E., Halsall, H.B. and Heineman, W.R. (1988) p-Aminophenyl phosphate: an improved substrate for electrochemical enzyme immunoassay. *Anal. Chim. Acta* 214:187–195.

Vorlop, K.D., Muscat, A. and Beyersdorf, J. (1992) Entrapment of microbial cells within polyurethane hydrogel beads with the advantage of low toxicity. *Biotechnol. Tech.* 6:483–488.

VROM/NL (1988) Leidraad Bodensanering, Deel II, Technisch-Inhondelijk Deel. In deutscher Übersetzung (1989): Leitfaden Bodensanierung, BMU, Teil 2.

Wehmeyer, K.R., Halsall, H.B. and Heineman, W.R. (1985) Heterogeneous enzyme immunoassay with electrochemical detection: competitive and "sandwich"-type immunoassays. *Clin. Chem.* 31:1546–1549.

White, S.F., Turner, A.P.F., Bilitewski, U., Schmid, R.D. and Bradley, J. (1994) Lactate, glutamate and glutamine biosensors based on rhodinised carbon electrodes. *Anal. Chim. Acta* 295:243–251.

White, S.F., Turner, A.P.F., Bilitewski, U., Bradley, J. and Schmid, R.D. (1995) On-line monitoring of glucose, glutamate and glutamine during mammalian cell cultivations. *Biosens. Bioelectron.* 10:543–551.

Wilmer, M., Renneberg, R. and Spener, F. (1996a) Rapid enzyme-immunoassay for the detection of 2,4-dichlorophenoxyacetic acid (2,4-D) in water using monoclonal antibodies. *Vom Wasser* 86:83–93.

Wilmer, M., Trau, D., Renneberg, R. and Spener, F. (1996b) Amperometric immunosensor for the detection of 2,4-dichlorophenoxyacetic acid (2,4-D) in water. *Anal. Lett.*, in press.

Xie, B., Mecklenburg, M., Danielsson, B., Öhman, O., Norlin, P. and Winquist, F. (1995) Development of an integrated thermal biosensor for the simultaneous determination of multiple analytes. *Analyst* 120:155–160.

Frontiers in Biosensorics II
Practical Applications
ed. by. F.W. Scheller, F. Schubert and J. Fedrowitz

Enzymatic substrate recycling electrodes

U. Wollenberger, F. Lisdat and F.W. Scheller

University of Potsdam, Insitute of Biochemistry and Molecular Physiology, c/o Max-Delbrück-Center for Molecular Medicine, D-13122 Berlin, Germany

Summary. A weak chemical signal might result in a large response when biochemically amplified. Enzymatic recycling of the analyte is one of the biochemical ways of providing an effective increase in biosensor sensitivity by several orders of magnitude. The enhancement of sensitivity is provided by consecutive consumption and generation of the analyte on the sensor surface. The principle of enzymatic substrate regeneration using bioelectrocatalysis and coupled enzymes is shortly reviewed and illustrated with some recent developments of biosensors for catecholamines, and its potential for electrochemical immunoassays is outlined.

Introduction

Sensitivity is one of the key features of a biosensor. It has to be optimized in order to make the sensor applicable for the particular target analyte. The relevant measuring range may extend over 10 decades. In some cases, e.g., when fermentation products are to be measured, the concentrations are in the lower molar range. Metabolites in the blood and urine, such as glucose, uric acid and lactate, may be present in micro- to millimolar concentration. Other compounds, such as hormones, drugs and signal transmitters in physiological fluids, xenobiotics, organic pollutants, pesticides etc. in the environment and in food products, appear only in minute amounts. In these cases the concentration is too low to be measured with enzyme sensors.

Electroanalytical techniques are fairly sensitive and currents as low as 10^{-10} A can be recorded with commercial devices. Thus under optimum conditions – with high enzyme loading under fast mass transport in thin layers and efficient external mass transfer – enzyme electrodes can measure substrates down to 10^{-6} mol/l with acceptable precision. For the measurement of substrates in the nanomolar range an increase of sensitivity of the enzyme electrode is required. One way to solve this problem is the continuous regeneration of the analyte in cyclic reactions.

The combination of the electrochemical detection principle and the recycling of the analyte can be performed in a number of ways (Fig. 1).

In the electrochemical (Niwa et al., 1991, Wollenberger et al., 1994) and bioelectrocatalytic (Scheller et al., 1987; Ortega et al., 1993) approach the target analyte is recycled between electrodes or electrode and redox centre of the enzyme and is therefore necessarily limited to reversible redox species. In contrast, bienzymatic substrate recycling (Schubert et al., 1985;

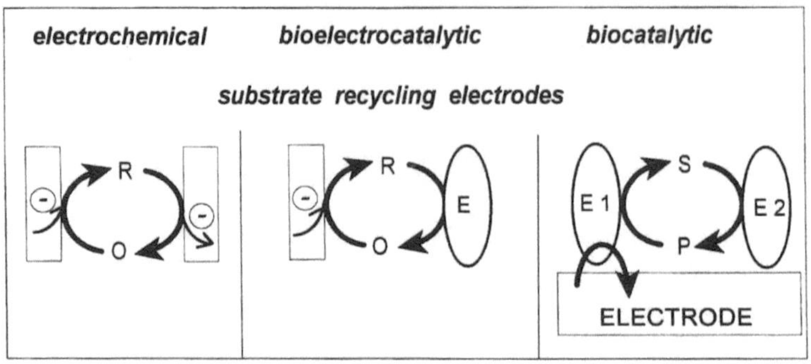

O - oxidized species
R - reduced species
S - substrate
P - product
E_n- enzyme

Figure 1. Illustration of the principles of substrate recycling electrodes.

1990a; Wollenberger et al., 1993) may be performed between enzymes, where one enzyme forms the substrate of the other, while a coreactant is measured directly or in an additional analytical step.

The present contribution gives an overview over the present state of enzyme-based recycling systems for increase of sensitivity of enzyme electrodes and illustrates examples for the application both for substrate analysis and enzyme immunoassays.

Bi-enzymatic substrate recycling

In the bienzymatic approach, the sensitivity enhancement is provided by shuttling the analyte between enzymes acting in cyclic series of reactions accompanied by cosubstrate consumption and accumulation of by-products.

The target analyte can be substrate or coenzyme of the respective enzyme. Assuming a sufficiently high activity of one enzyme in the presence of its cosubstrate and an analyte at a concentration far below its Michaelis constant, an amplification is achieved by turning on the second enzyme by addition of its cosubstrate. By measuring the concentration change of one of these coreactants directly or in an additional analytical step, the recycling system is used as a biochemical amplifier for the analyte, which can be each of the substrates of the participating enzymes.

This principle of signal amplification was first successfully applied to a number of systems involving the indication of pyridine dinucleotides using soluble enzymes (Lowry and Passonneau, 1972). For soluble enzymes at steady state the overall cycling constant *k,* which represents the amplification, is only dependent on the apparent first order rate constant k_i of the enzymes used:

$$k = k_1 k_2/(k_1 + k_2) \tag{1}$$

For enzymes in solution in some cases rates of 20,000 per hour can be obtained. With double recycling total amplification of 4×10^8 has been achieved (Lowry and Passonneau, 1972). For enzyme electrodes, additionally, diffusion has to be considered. According to theory (Kulys et al., 1986) the amplification factor, G, is under steady state conditions:

$$G = k_1 k_2 L^2/2D(k_1 + k_2) \tag{2}$$

where L is the membrane thickness, and D the diffusion coefficient. Obviously, at high activities of both enzymes immobilized into the enzyme layer with high characteristic diffusion time (L^2/D), the possible amplification is very large, too.

When one molecule of product is formed per substrate molecule the total concentration of intermediate substrates remains constant and the concentrations of the coreactants increase or decrease linearly with time. Then the number of cycles in which the substrate is turned over in a given time is a function of the substrate concentration.

A tremendous signal amplification is expected if in the cycling reaction more than one analyte molecule is regenerated. Here the total amount of intermediates and by-products increases exponentially with time. Theoretical considerations have shown that the concentration of any of the cycling intermediates or byproducts at any given time is a linear function of the initial substrate concentration (Kopp and Miech, 1972). An example illustrating this principle is the ADP/ATP cycling system using myokinase/pyruvate kinase (Pfeiffer et al., 1994). With an optimized sensor configuration an increase of sensitivity for ADP by a factor of 800 was obtained, resulting in a measuring range between 50 nmol/l and 2 µmol/l.

The concept of linear enzymatic signal amplification was realized by coupling dehydrogenases with oxidases or transaminases, phosphorylases and phosphatases, or by coupling kinases with each other (Table 1). When oxidases are involved, electrode-detectable species are included in the reaction scheme. Therefore, the change of coreactant concentration can be measured directly at the electrode onto which the recycling enzyme pair is immobilized. As can be seen in Table 1, in most cases oxygen consumption has been measured, but also potential and pH changes have been indicated. The recycling systems based on kinases and phosphorylases/phosphatase were combined with indicator enzymes in order to form an electrochemically detectable species.

Depending on the measurement arrangement, enzymes, and membrane materials used, amplification by several orders of magnitude has been rea-

Table 1. Enzymatic substrate recycling electrodes

Analyte	Enzyme couple	Transducer	Amplification	Comment	Reference
ADP	myokinase/pyruvate kinase	oxygen electrode with pyruvate oxidase	800	exponential cycle	Pfeiffer et al., 1994
ADP/ATP	hexokinase/pyruvate kinase	oxygen electrode with lactate DH/lactate monooxygenase	220		Wollenberger et al., 1987b
		modified carbon electrode with glucose-6-phosphate DH	1200	with intermediate accumulation	Yang et al., 1991
Adrenaline p-aminophenol	oligosaccharide DH/laccase	oxygen electrode	3000	for enzyme immunoassay [1]	Bier and Scheller 1996a
Adrenaline phenol derivatives	(PQQ)glucose DH/ tyrosinase	oxygen electrode	100–1000	determination of catecholamines; phenols in water; alkaline phosphatase in ELISA [2]	Makower et al., 1995
Adrenaline, p-aminophenol	(PQQ)glucose DH/laccase	oxygen electrode	5000 10000	also ferrocenes, for redox label and enzyme [3] immunoassay	Ghindilis et al., 1995a Ghindilis et al., 1995b
		potentiometric redox electrode	200	potentiometric bio-electrocatalytic detection	Ghindilis et al., 1995c
		antimony pH electrode	1000		Eremenko et al., 1995b
		pH-FET			Pfeiffer et al., 1996

Benzoquinone/ hydroquinone	fructose DH/laccase	oxygen electrode	700		Wollenberger unpublished
Benzoquinone/ hydroquinone	cytochrome b2/laccase	oxygen electrode	500		Scheller et al., 1992
Ethanol	alcohol oxidase/ alcohol DH	oxygen electrode			Hopkins, 1985
Glucose	glucose oxidase/ glucose DH	oxygen electrode	10		Schubert et al., 1985
Glutamate	glutamate DH/ alanine aminotransferase	modified carbon electrode	15		Schubert et al., 1986
Glutamate	glutamate oxidase/ glutamate DH	oxygen electrode	20		Wollenberger et al., 1989
Glutamate	glutamate oxidase/ alanine aminotransferase	hydrogen peroxide electrode	500		Yao et al., 1989
L-Leucine	leucine DH/ amino acid oxidase	oxygen electrode	40		Scheller et al., 1990
Lactate/pyruvate	cytochrome b2/ lactate DH	Pt-electrode	10		Schubert et al., 1985
		glassy carbon electrode			Vidziunaite and Kulys, 1985
Lactate/pyruvate	lactate oxidase/ lactate DH	oxygen electrode	48000	used for enzyme immunoassay[4];	Scheller et al., 1992
			4100	sialinic acid determination.[5]	Wollenberger et al., 1987a
			250		Mizutani et al., 1985
		Pt-ring electrode	10^9	rotating dual reactor electrode	Raba and Mottola, 1995
Malate/oxalacetate	malate DH/ lactate monooxygenase	oxygen electrode	3		Scheller et al., 1988

Table 1 (continued)

Analyte	Enzyme couple	Transducer	Amplification	Comment	Reference
$NADH/NAD^+$	peroxidase/glucose DH	oxygen electrode	60		Schubert et al., 1985
$NADH/NAD^+$	diaphorase/glycerol DH	glassy carbon electrode	800–1200		Tang and Johannson, 1995
$NADH/NAD^+$	NADH oxidase/alcohol dehydrogenase	oxygen electrode	100	enzyme pair used in conjunction with H_2O_2 detector for enzyme immunoassays[6]	Mizutani et al., 1993
$NADPH/NADP^+$	NADPH oxidase/glucose-6-phosphate DH	oxygen electrode	50		ibid
Peptides (containing tyrosine)	(PQQ)glucose DH/ tyrosinase	oxygen electrode			Eremenko et al., 1996
Phosphate	nucleoside phosphorylase/ alkaline phosphatase	oxygen electrode with xanthine oxidase	20		Wollenberger et al., 1992
Phosphate	maltose phosphorylase/ acid phosphatase	hydrogen peroxide electrode with glucose oxidase	15		Conrath et al., 1995

[1] Bier et al., 1996; [2] Bauer et al., 1996; [3] Scheller et al., 1995; [4] Makower et al., 1994; [5] Pfeiffer et al., 1996; [6] Athey and McNeil, 1994.

lized. Only for those sensors where high activity of the both enzymes is applied can a high amplification be obtained. When dealing with extremely high amplification one has to bear in mind, however, that the sensor signal becomes highly susceptible to minute amounts of contaminants affecting the enzyme reactions.

So far, the highest amplification was obtained for pyruvate and lactate determination by using the lactate oxidase/lactate dehydrogenase pair. The oxygen consumption in the gelatin membrane bearing both enzymes is enhanced, yielding an increase in the sensitivity to lactate by factor of up to 4100 (Wollenberger et al., 1987a). The experimental results have been found to be in line with theory (Equation 2). Thus, the study of the influence of the enzyme loading on the amplification of the lactate signal showed that a 4100-fold amplification was reasonable using the gelatin matrix with the characteristic diffusion time of about 90 s. With this sensor lactate concentrations as low as 1 nmol/l could be determined with acceptable precision. When the immobilization was performed with polyurethane which permits higher enzyme loading, up to 48 000-fold enhancement for the lactate response was achieved (Scheller et al., 1992). However, the reproducibility and stability of this large amplification is poor. The amplification of the lactate response decreases with progressive enzyme inactivation during sensor operation, while the unamplified lactate signal (using only lactate oxidase) remains stable. This behaviour is due to the fact that the enzyme excess maintains diffusion control of the sensor for the simple process. Amplification can be tuned by limiting the cosubstrate. Thus, employing the recycling sensor 1–2 orders above the possible detection limit more stable response is obtained and allows for the combination with a displacement enzyme immunoassay for the determination of cocaine (Makower et al., 1994). Furthermore the sensor has been applied to the measurement of sialinic acid (Pfeiffer et al., 1996).

Sensors with high sensitivity can be also constructed on the basis of phenol-oxidizing enzymes, which are currently available with high catalytic activity. For example phenol oxidases, i.e. laccase and tyrosinase, which oxidize a wide range of substances including catecholamines, phenols, and redox dyes by dissolved oxygen (see Peter and Wollenberger, this volume), have been used in combination with heme and pyrroloquinoline quinone (PQQ) containing (NAD(P)H independent) dehydrogenases. The highest sensitivity was obtained for aminophenol (subnanomolar concentration) when quinoprotein glucose dehydrogenase from *Acinetobacter calcoaceticus* and laccase from *Coriolus hirsutus* were coentrapped (Ghindilis et al., 1995a). Because of its group specificity, the electrode can also be used for the detection of low amounts of redox dyes (Ghindilis et al., 1995b), which is of particular interest for the development of immunoassays (see below).

Similar sensitivity is obtained when mushroom tyrosinase substitutes laccase. The change in specificity makes this arrangement suitable for sensitive determination of phenols in aqueous samples (Makower et al.,

1995) and tyrosine containing peptides (Eremenko et al., 1996). The application of these sensors for measurement of catecholamines is discussed below. Other acceptor dependent dehydrogenases such as fructose dehydrogenase and oligosaccharide: acceptor oxidoreduct (oligosaccharide dehydrogenase) have been employed in place of the quinoprotein glucose dehydrogenase. When using the latter enzyme a rather broad spectrum of substances, such as aminophenols, diamines, and catecholamines is accessible to sensitive detection, but with different sensitivities as compared with the laccase-glucose dehydrogenase couple (Bier and Scheller, 1996).

Recycling systems are not necessarily limited to reactions in which electrochemically active compounds are involved. In these other cases the recycling enzyme pair is combined with an indicator enzyme (or sequence) transforming one of the cycle coreactants (mostly a product) into a measurable species. Owing to their usually favourable equilibrium constants kinase reactions are well applicable to such recycling experiments (Wollenberger et al., 1987b; Yang et al., 1991).

As can be seen from Table 1, for the sensitive measurement of ADP and ATP biosensors with various indicator reactions have been constructed, for example, by combining an oxygen electrode and a gelatin layer comprising the pair pyruvate kinase and hexokinase and a lactate dehydrogenase and lactate monooxygenase sequence. The latter sequence transforms the pyruvate formed in the recycling reaction into an oxygen consumption signal (Wollenberger et al., 1987b). In contrast to lactate/pyruvate recycling sensors ADP and ATP were indicated with different sensitivities, which were mainly attributed to their deviating diffusion coefficients. The sensitivity in the linear measuring range for ADP was amplified by a factor of 220, whereas ATP response was slightly higher. Other indicator reactions used were pyruvate oxidase and glucose-6-phosphate dehydrogenase (Yang et al., 1991).

Two different cycling enzyme arrangements have been proposed for the determination of inorganic phosphate. Interestingly, both setups include the production of two indicator molecules for each cycle of phosphate conversion. The first approach comprises nucleoside phosphorylase and alkaline phosphatase for phosphate recycling, and xanthine oxidase for indication of the liberated hypoxanthine (Wollenberger et al., 1992). Hypoxanthine is oxidized in a two-step reaction; thus, per molecule of phosphate cycled or hypoxanthine formed, two molecules of oxygen are consumed. The phosphate-dependent phosphorylation of maltose by maltose phosphorylase proceeds with formation of one molecule of glucose and glucose-1-phosphate each. The hydrolysis of the latter by acid phosphatase liberates a second glucose molecule and phosphate, which enters the cycle again. A glucose oxidase membrane was added in order to indicate glucose formation (Conrath et al., 1995).

Bioelectrocatalytic substrate recycling

One enzyme of the enzymes may be substituted by a redox electrode. Then the enzymatic reaction is followed by an electrochemical regeneration (or vice versa) resulting in a catalytic current, which is the measured signal. As mentioned above, this type of substrate recycling is limited to reversible redox species.

Already in 1984 Wasa reported on the enhancement by laccase of the sensor response for noradrenaline, hydroquinone, and p-phenylene diamine (Wasa et al., 1984). Since then a number of other enzymes have been investigated, which belong in the great majority of cases to the family of copper oxidases. But also flavoenzymes, heme-containing enzymes and PQQ-dependent dehydrogenases have been used (Table 2).

In ideal cases of bioelectrocatalytic recycling the potential required for regeneration is low (to avoid electrochemical interference), sufficient (co)substrate of the enzyme is present (in order not to limit the reaction), and the analyte is stable in both redox states.

Vital for a rapid heterogeneous electron transfer is the close contact of enzyme and electrode material. Therefore surface immobilization using adsorption, covalent binding, and entrapment of the redox enzyme, and bulk modification procedures have been established, the latter appearing to be the most effective way. Improvement results from the additional integration of modifier and promoter, which bind the protein to the electrode surface and, while not itself taking part in the electron transfer process, encourage electron transfer with the protein to proceed (Gorton, 1995). Recent developments profit from modern fabrication technologies by combining enzymes with screen-printed electrodes (Wang and Chen, 1995 b) for remote sensors and microfabricated interdigitated gold electrodes for gas phase application (Dennison et al., 1995).

The substances measured so far with the bioelectrocatalytic approach are redox mediators, electroactive metabolites, phenolics and drugs. Enzymes liberating those compounds from electroinactive precursors (in the potential region investigated) may be quantified, too.

The highest sensitivity may be obtained when both the electrode and enzyme reaction are very fast and the intermediate substances (oxidized and reduced form of the analyte) are stable. As can be seen from Table 2, in most reported cases bioelectrocatalysis has been accomplished with carbonaceous electrode material. This material efficiently facilitates fast redox conversion of quinones and related compounds. For example, the quinoprotein glucose dehydrogenase, tyrosinase and laccase preparations are of very high enzyme activity. Therefore the combination of these enzyme with carbonaceous electrodes results in ultra-sensitive sensors with detection limits even below 1 nmol/l. At the same time the regeneration prevents fouling of the electrode surface by polymeric products and therefore induces stability. Remarkable short response times (a few seconds) are achieved

Table 2. Bioelectrocatalytic sensors for selected phenolic compounds[1]

Enzyme	Electrode material	Substrate[1]	Detection limit, nmol/l[1]	Comment	Reference
Cytochrome b_2	glassy carbon		500	gelatin membrane, + lactate	Scheller and Schubert, 1989
Diaphorase	glassy carbon	p-aminophenol	0.5	aP determination; diaphorase in solution, + NADH	Yamaguchi et al., 1992
Fructose dehydrogenase	glassy carbon			+ fructose	Ikeda et al., 1991
(PQQ)glucose dehydrogenase	glassy carbon	p-aminophenol	0.5–20	PVA, + glucose	Eremenko et al., 1995a
	carbon paste	p-aminophenol	0.5	bulk modification with PEI promoter, + glucose	Wollenberger et al., 1995
	gold	p-aminophenol	5.0	covalent multi-layer assembly, + glucose	Jin et al., 1995
Glucose oxidase	glassy carbon	hydroquinone/ catechol	1.0	photocrosslinked PVA-stilbazolium, + glucose	Mizutani et al., 1991
	edge-plain pyrolytic graphite		100	PVA/glutardialdehyde + glucose	Moore et al., 1994
	graphite	acetaminophen		+ glucose	Ikeda, 1984
Oligosaccharide dehydrogenase	carbon paste	p-aminophenol	5	+ glucose	author
	carbon ink	p-aminophenol	50	photopolymer, + glucose	
Peroxidase (HRP)	glassy carbon	2-amino-4-chlorophenol	20	adsorption, halogenated phenols + H_2O_2,	Ruzgas et al., 1995
	carbon paste			bulk modification, with lactitol	
Laccase	reticulated vitreous carbon	hydroquinone p-phenylene diamine	70	glutardialdehyde	Wasa et al., 1984

	pyrographite			gelatin	Scheller et al., 1987
	glassy carbon	hydroquinone	600	AQ, measurements in organic solvents	Wang et al., 1993b
	carbon paste	p-aminophenol	2		
	carbon ink	p-aminophenol		photopolymer	author
	carbon based electrodes	catechol		coimmobilized tyrosinase	Yaropolov et al., 1995
Tyrosinase	carbon paste, glassy carbon, graphite	catechol/phenol	10/13	fixed on surface with dialysis membrane	Skladal, 1991
	glassy carbon	phenol	100	AQ, organics accumulation	Wang et al., 1993a Wang and Chen, 1995a
	carbon paste	phenol	3	bulk modification with PEI promoter	Ortega et al., 1993
	graphite	catechol	2	adsorption, carbodiimide	Ortega et al., 1994
	carbon paste	catechol/phenol	40/1000	graphite epoxy	Önnerfjord et al., 1995
	carbon paste	phenol	10	graphite modified with PTFE (gaspermeable electrode)	Kaisheva et al., 1996
	carbon ink	catechol	100	thick-film electrode	Wang and Chen, 1995b
	carbon	phenol	1000	adsorbed, measurement in chloroform	Hall et al., 1988
	gold	phenol		in glycerol-electrolyte on interdigitated electrode, gasphase sensor	Dennison et al., 1995

[1] substrate which was measured with the highest sensitivity.

when the enzyme is covalently bond to an activated self-assembled monolayer on a gold surface (Jin et al., 1995). Furthermore, this binding procedure prolonged considerably the working stability of quinoprotein glucose dehydrogenase modified sensors as compared to bulk modified (Wollenberger et al., 1995) and polymer membrane (Eremenko et al., 1995a) detectors. With intermittent PQQ incubations the electrodes could be used for more than 4 weeks.

Substituting the electrode by a reducing agent may also improve sensitivity due to a regeneration process (Macholan, 1990; Uchiyama et al., 1993). For example, the addition of ascorbic acid to a phenol oxidase modified electrode improves the sensitivity by a factor of 300 (Uchiyama et al., 1993). The rate of the chemical regeneration, i.e. recycling efficiency, is lower than the electrode reaction and therefore the sensors are not as sensitive as the bioelectrocatalytic systems.

Application of enzymatic recycling systems

Due to the high sensitivity the recycling sensors have potential for application both for substrate measurement and electrochemical immunoassays. Whereas the basic principles of enzymatic substrate regeneration in biosensors has been illustrated for a number of substrates during the last 10 years, only recently has substantial work focused on their application.

Sensitive detection of catecholamines

During the last decade the sensitive determination of catecholamines has become a favorite research target. Studies have concentrated on the quantification of their plasma content as well as their release by cell cultures and single cells. In the medical field the improved prognosis of tumoric diseases and heart failure resulted in significant progress in HPLC analysis using electrochemical or optical detection (e.g. Chen et al., 1994; Liu and Wang, 1994). For the study of the catecholamine content in or secretion by cells, modified microelectrodes and capillary electrophoresis in combination with fluorescence or electrochemical detection were preferably used (Ewing et al., 1992; Jankowsky et al., 1995; Wightman et al., 1995). To illustrate the relevant concentration range it shall be mentioned that the daily catecholamine excretion in the urine of a healthy person is around 5 μmol/d, whereas the plasma concentration does not exceed 10 nmol/l. Recycling systems have proved to be principally suited to this analytical range. Therefore considerable attention was given to catecholamine sensor development. From the previously studied bioelectrocatalytic amplification systems detection limits of 100 nmol/l adrenaline (Wasa et al., 1984) or

Table 3. Comparison of different amplification systems for the determination of catecholamines

Redox couple	$t_{90\%}$	Prefered substrate	Limit of detection
a) monoenzymatic			
GCE/GDH [1,2]	3–4 min	noradrenaline dopamine	1 nmol/l
Laccase/GCE [1]	2–3 min	dopamine	10 nmol/l
CPE/GDH [2,3,4]	1–2 min	adrenaline	1 nmol/l
Laccase/CPE [3,4]	2–3 min	noradrenaline	10 nmol/l
b) bienzymatic			
Laccase/GDH [2]	1–2 min	adrenaline	0.5 nmol/l
Tyrosinase/GDH [2]	2 min	dopamine	25 nmol/l

[1] GCE – glassy carbon electrode; [2] GDH – quinoprotein glucose dehydrogenase; [3] CPE – carbon paste electrode; [4] Dopamine has not been investigated.

300 nmol/noradrenaline (Moore et al., 1994) have been reported. Our intention is directed to the ultrasensitive detection of the neurotransmitters below the concentration of 10 nmol/l.

The catecholamines, adrenaline, noradrenaline and dopamine are redox active substances, which can be easily oxidised and, in their quinone form, easily reduced. Furthermore they can serve as electron donors for copper-containing oxidases like tyrosinase (EC1.14.18.1), ceruloplasmin and laccase (EC1.10.3.2) and in their oxidised form as electron acceptors for enzymes such as quinoprotein glucose dehydrogenase (GDH, EC1.1.99.17) or oligosaccharide dehydrogenase (ODH) (Bier and Scheller, 1996). Therefore several combinations of redox couples were studied with respect to their potential for practical catecholamine analysis. Table 3 gives an overview of the systems under detailed investigation.

Bioelectrocatalytic amplification

Two principal ways can be gone in the monoenzymatic approach using the electrode for either oxidation or the reduction process. For the realisation of the first way a PQQ-dependent glucose dehydrogenase was fixed onto a glassy carbon electrode (GCE) held at a potential of +0.5 V vs Ag/AgCl, which is sufficient for effective catecholamine oxidation (Eremenko et al., 1995a). The ability of quinoprotein glucose dehydrogenase to reduce the quinone product is provided in the presence of glucose which is converted to gluconolactone. The advantage of this system is that it works independent of the oxygen content of the solution. For immobilisation a PVA-matrix was used.

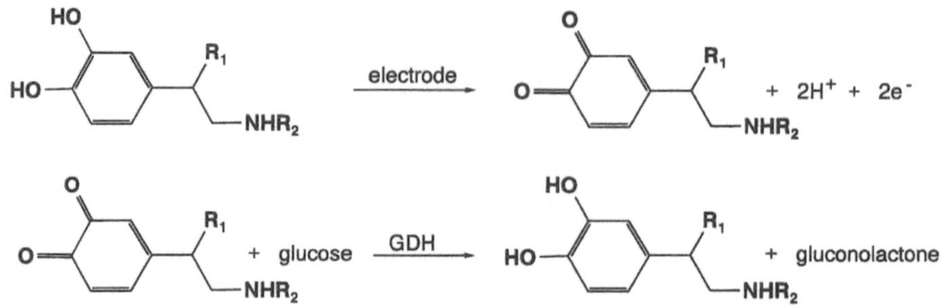

	R_1	R_2
adrenaline	OH	CH_3
noradrenaline	OH	H
dopamine	H	H

The quinoprotein glucose dehydrogenase – modified glassy carbon electrode shows very sensitive amperometric response to catecholamines, with noradrenaline as the preferred substrate. The linear region extends from 2 to 300 nmol/l. By variation of the enzyme amount in the polymer membrane it could be demonstrated that the sensitivity is increased with increasing enzyme content. At very high enzyme loading (800 U/cm^2) even the selectivity pattern is changed: the three catecholamines give nearly the same response. This suggests that the greater tendency of the quinonoid product from adrenaline for intramolecular cyclization (Matysik et al., 1992; Ciolkowski et al., 1994) is the main reason for the selectivity pattern obtained (response: noradrenaline ~ dopamine > adrenaline).

A sensor combining laccase with a glassy carbon electrode poised at – 0.25 V vs Ag/AgCl (in order to avoid oxygen interference) responds to the three catecholamines in the range 20–300 nmol/l but with different sensitivity. The lower limit of detection (signal/noise ratio >3) is around 10 nmol/l for the most sensitively registered analyte dopamine. Whereas for urine analysis this detection limit is adequate, for plasma analysis the amplification system suffers from a lack of sensitivity.

For the amplification systems the close contact between the oxidising and the reducing partner is of great importance in ensuring fast transport of the shuttled molecules. By embedding the enzymes in a carbon paste electrode (CPE) very small diffusion distances can be provided. On the other hand the environment of the enzyme is completely changed compared to the aqueous membrane. This results in a changed selectivity pattern compared to GCE. Whereas, for example, the GCE/quinoprotein glucose dehydrogenase sensor shows the following graduation in response: Noradrenaline ≥ dopamine > adrenaline, the CPE/quinoprotein glucose dehydrogenase electrode gave the opposite order: Adrenaline > noradrenaline. The sensitivity as well as the selectivity of carbon paste electrodes are a complex matter with significant influence by the electrode material, pretreatment procedures and paste additives.

Biocatalytic amplification

According to the theory of the bienzymatic approach increasing enzyme activity directly influences the sensitivity of the system. Therefore advantage was taken by the use of highly active enzymes like laccase from *Coriolus hirsutus* (320 U/mg), mushroom tyrosinase (~ 4000 SigmaU/mg) and quinoprotein glucose dehydrogenase from *Acinetobacter calcoaceticus* (350–800 U/mg).

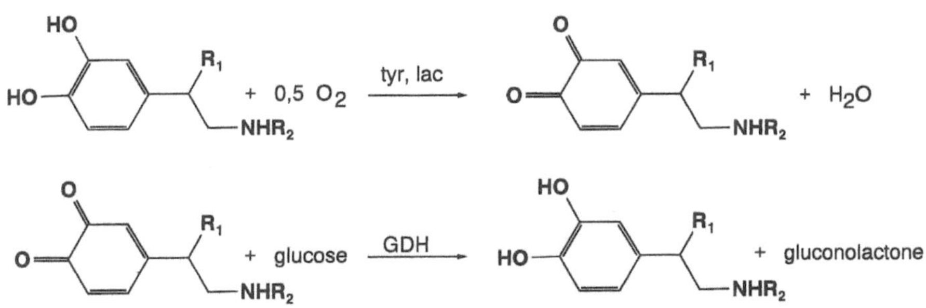

By combining laccase with glucose dehydrogenase (Eremenko et al., 1995 b, Ghindilis et al., 1995 a) even subnanomolar concentrations could be detected with adrenaline as the favoured substrate (Fig. 2). The reversibility of the sensor response could be shown in flow measurements. The response time ($t_{90\%}$) is in the range of 60–100 s. This remarkable result could be obtained despite the different pH optima for the enzymes of the couple (laccase pH 4, quinoprotein glucose dehydrogenase pH ~7). Measurements were performed at pH 6 in phosphate buffer at 25°C. The bienzymatic sensors allow the detection of catecholamines over a period of around 10 days. During this time the sensor response is decreased to 50–60% of the initial value.

The high recycling efficiency of the system is caused by the high enzyme activity as well as the low shuttling distances between the two enzymes. Compared to the analyte concentration to be detected (~ 10 nmol/l) the enzyme concentration within the membrane is around 4 orders of magnitude higher, so that the internal transport between the two reaction sites becomes very fast. If the lower limit of detection of the amplification sensor (0.5 nmol/l adrenaline) is compared to the unamplified sensor (only laccase activity: 2.5 µmol/l) it can be clearly seen that the laccase reaction

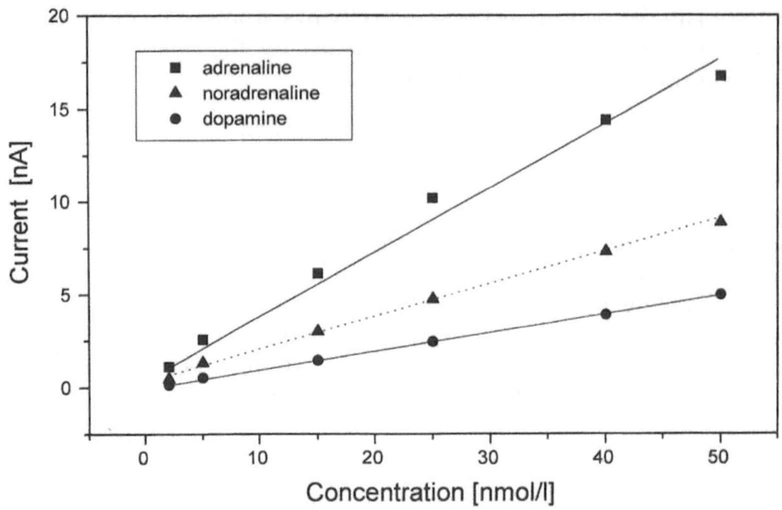

Figure 2. Sensitivity of the laccase/quinoproteinglucose dehydrogenase electrode for the different catecholamines.

is almost 5000 times amplified by this bienzymatic combination (Fig. 3). For the glucose consumption follows that for the detection of one molecule of the neurotransmitter about 5000 glucose molecules are oxidised. Therefore a relatively high glucose concentration in solution has to be adjusted (10–25 mmol/l).

The tyrosinase/quinoprotein glucose dehydrogenase sensor appears to be less sensitive. The lower detection limit is 25 nmol/l, with dopamine as the most sensitive catecholamine. In contrast, adrenaline is registered as about 5 times less sensitive. The response rate is quite similar to the laccase/quinoprotein glucose dehydrogenase sensor.

The selectivity of each amplification sensor is determined by different factors, such as substrate specifity of the enzyme, diffusion properties of the analyte in the immobilisation matrix, affinity of reaction products to the matrix, and stability of the oxidised or reduced catecholamine (side reactions). These factors might be advantageously used to obtain different sensing characteristics for each system. As for the bioelectrocatalytic approach, for the bienzymatic sensors enzyme loading appears as a valuable control parameter. All sensors respond to the sum of the catecholamines, therefore the combination of sensors with a different and defined selectivity offers the possibility to evaluate the individual catecholamine concentrations by pattern recognition.

Amplification in the bienzymatic approach is not only influenced by the total enzyme activity, but by the relation of both enzyme partners too. Loading of the PVA membrane with increasing tyrosinase content clearly demonstrates the possibility to improve the sensitivity. An increase in the

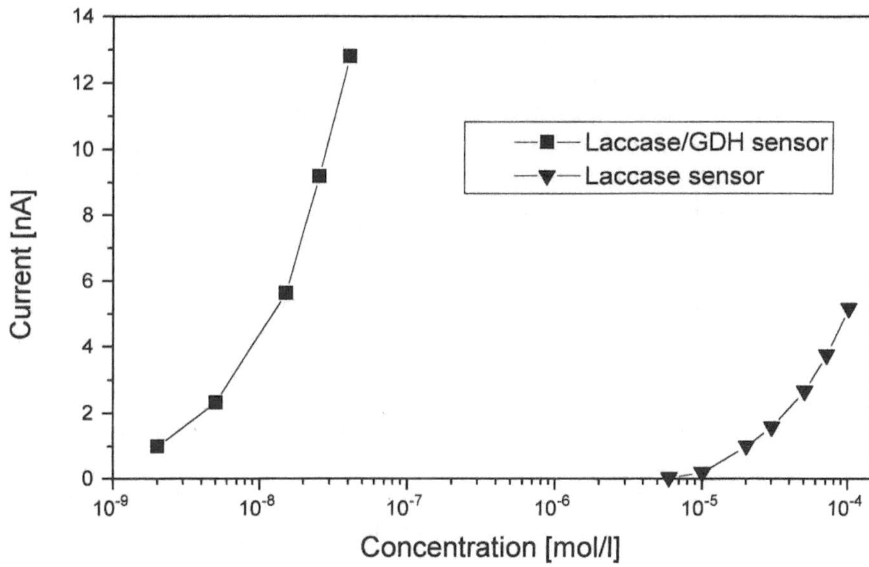

Figure 3. Response of the amplified and unamplified laccase sensor to adrenaline.

tyrosinase content of the membrane from 40 U/cm^2 to 200 U/cm^2 resulted in an increase of the sensor response by one order of magnitude. However, for practical sensor fabrication a lower reproducibility in the membrane preparation has to be taken into consideration at high enzyme loadings.

Interference problems

The most common interfering substances in physiologial media are reducing compounds such as ascorbic or uric acid. Therefore the sensor systems were investigated with respect to their interferant response. The laccase/quinoprotein glucose dehydrogenase electrode responds to ascorbic acid with a 1000 times lower sensitivity than to adrenaline. In addition, the response to the catecholamines is only slightly influenced in the presence of an ascorbic acid excess by a factor of 1000–10000. In contrast uric acid surpresses the catecholamine sensitivity and appears as a substrate of the sensor. Intermediates of the uric acid conversion inhibit the enzyme couple. This can be concluded from the decrease of the measuring signal with time after the response to uric acid. The effect increases with rising concentration. In conclusion it is indicated that both substances should be removed from the sample by an appropriate enzyme, as was shown by incubation of the sample with ascorbate oxidase.

On the other hand there are several fields of application where these substances are absent and other matrices have to be considered. For example, the secretion of catecholamines by chromaffine cells could be successfully monitored by use of the laccase/quinoprotein glucose dehydrogenase sensor (Ghindilis et al., 1995 b). Chromaffine cells respond to stimuli, such as nicotine, by secretion of noradrenaline, adrenaline and traces of dopamine. The study of such cell reactions is of importance for a better understanding of signal processing in biological systems. By comparison of the catecholamine detection using the bienzymatic electrode with a conventional HPLC analysis it becomes evident that the biosensor is able to determine the catecholamine secretion even in such a complex matrix of agonist, buffers and co-excreted hormones. No interference effects could be observed.

A further interesting field of application which is presently under investigation is the measurement of catecholamines in reperfusion experiments on a double heart model. In the effluent, which passed through an isolated animal heart, adrenaline determination is possible after adjusting the pH with an appropriate buffer solution. Other metabolites caused no detectable response. This might offer prospects in investigating medically relevant effects caused by the infarction itself or the following reperfusion of the heart.

Recycling systems that shuttle the target analyte between two redox partners, seem to be an effective alternative to methods of instrumental analytical chemistry in determining the neurotransmitters adrenaline, noradrenaline and dopamine. For the nanomolar analysis the laccase/quinoprotein glucose dehydrogenase- and the GCE/quinoprotein glucose dehydrogenase-electrodes proved to be the most appropriate candidates. Because of the rather broad substrate spectrum the enzyme electrodes cannot distinguish between the three catecholamines, therefore the sum of all is recorded. Here the different selectivity of each system offers the possibility to obtain detailed analytical information by pattern recognition. Finally it should be noted that a fruitful combination of separation steps by HPLC and specific recognition by an amplification sensor may open further fileds of application.

Immunoassays using enzymatic substrate recycling electrodes

The combination of immunoreactions and electrode-based substrate recycling connects very specific recognition of an analyte with highly sensitive detection. Most important for this field of application is the sensitivity, which permits to detect a label at very low concentration. For enzyme immunoassays in particular, several goals have to be achieved, (i) enzyme substrate produce only a low background response, (ii) substrate and product are stable, (iii) enzyme had high turn-over number, (iv) product is indicated very sensitively, (v) close pH optimum of marker enzyme and detector. Since the sensor is used to quantify the change of one particular substance the broad substrate specificity of the enzymes used is not critical.

Therefore sensors using acceptor-dependent dehydrogenases in combination with phenol oxidases (Ghindilis et al., 1995a; Bier et al., 1996; Bauer et al., 1996) or carbon paste electrodes and the lactate oxidase/lactate dehydrogenase couple (Makower et al., 1994) have been used. The enzymes measured so far are alkaline phosphatase and β-galactosidase (Scheller et al., 1995).

In the 'simplest' approach for the determination of antigens the recycling electrode can be used in combination with a conventional enzyme-linked immunosorbent assay (ELISA). In both sandwich-immunoassays and competitive immunoassays the electrode traces the enzyme marker, which generates the substrate for the recycling electrode from a non-detectable precursor.

For hapten determination the new dimension of sensitivity of enzyme electrodes permits a novel approach to competitive immunoassays where a substrate of the recycling electrode itself is used as marker of a hapten. The competition between the labelled hapten and the analyte hapten for binding to the respective antibody is indicated with the sensor. Since only the free (unbound) labelled hapten is measurable the whole assay can be performed in a homogeneous format. The indication of a displaced immobilized antibody-bound labelled hapten in a flow system is a further step towards automatic immunoassays.

Enzyme immunoassays

During the last years a number of electrochemical enzyme-immunoassays with alkaline phosphatase as label have been described (see McNeil et al., this volume). The formation of NAD^+ from $NADP^+$ (Athey and McNeil, 1994), pyruvate from phosphoenolpyruvate (Makower et al., 1994), phenol from phenyl phosphate (Bauer et al., 1996) and aminophenol from aminophenyl phosphate (Ghindilis et al., 1995b; Bier et al., 1996) have been followed using enzymatic substrate recycling and electrodes (see Table 1 and 2). Besides, β-galactosidase can also been employed; here an aminophenylated galactoside is hydrolysed. Determination of goat IgG and human tyroid stimulating hormone (hTSH) has been performed in a sandwich-type immunoassay using an oligosaccharide dehydrogenase/laccase electrode (Bier et al., 1996). In a first investigation alkaline phosphatase was used for the model compound IgG, and the liberation of p-aminophenol from p-aminophenyl phosphate was followed. This reaction, however, is known to suffer from drawbacks related to the limited stability of both p-aminophenyl phosphate and p-aminophenol in alkaline solution (Tang et al., 1988). Therefore a large blank signal, which in some cases exceeds the dynamic measuring range of the electrode, is obtained and the incubation time is limited. Furthermore, the use of alkaline phosphatase requires a change of the pH between the immunoassay and the electrode reaction.

Therefore two alternatives were studied – (i) β-galactosidase label and (ii) phenyl phosphate as substrate for alkaline phosphatase. Since the optimum pH of β-galactosidase is close to that of the bi-enzyme sensor (pH 6.5) the whole assay could be performed under the same conditions. The amount of bound β-galactosidase was traced after incubation with 1 mmol/l p-amino-phenyl-β-D-galactoside for 30 min. The sensitivity of the total assay was comparable to that of the photometric test.

For the determination of hTSH the sandwich-type assay was performed using biotinylated tracer antibody and streptavidin-β-galactosidase conjugate. The measuring range extends from 0.0005 to 20 ng/ml with a detection limit of 0.3 pg/ml; the c50-value is almost one order of magnitude lower than that of the photometric assay.

Bauer et al. (1996) used a tyrosinase and quinoprotein glucose dehydrogenase covered oxygen electrode as fast readout of a competitive immunoassay for the herbicide 2,4-D involving an alkaline phosphatase label where phenyl phosphate is used as substrate for alkaline phosphatase and the liberated phenol is indicated. When phenyl phosphate is a permanent additive of the buffer, the measurement can be performed free of substrate blanks. With this system 3.2 fmol/l alkaline phosphatase (320 zmol in a 100-µl sample) were detected after almost 1 h incubation with phenyl phosphate. The competitive immunoassay uses an immobilized antibody against 2,4-D in a microtiter plate. After competition between 2,4-D-alkaline phosphatase conjugates and 2,4-D for 1 h and a washing step, substrate buffer was added and transfered to a biosensor equipped FIA after further 2 min. This means a reduction of the incubation time from overnight, which is the time required for the standard optical method with p-nitrophenyl phosphate. The working range is ca 0.1–10 µg/l. The results are in good agreement with the optical reference method.

Cocaine was determined using conjugate displacement and a laccase/quinoprotein glucose dehydrogenase modified Clark-type oxygen electrode (Scheller et al., 1995). The displacement assay was performed in two ways. Monoclonal antibodies against cocaine were immobilized on a microtitre plate and cellulose beads (AvidGel AX) in a microcolumn saturated with alkaline phosphatase-cocaine conjugate. This displacement of the conjugate by cocaine is followed by indicating the enzyme in the supernatant (off-line) in the well or in the eluent of the column. Whereas a long incubation time permits 10 nmol/l–10 µmol/l cocaine to be detected with the flow system the detection limit is about 200 nmol/l (100 pmole).

Redox label immunoassay

The ability to detect nanomolar concentration of ferrocene derivatives with a quinoprotein glucose dehydrogenase/laccase sensor allows for the use of a ferrocene tracer in an immunoassay for low molecular weight antigens.

When an antibody is bound to the ferrocene-labelled hapten the redox label can not penetrate the dialysis membrane and is therefore not measurable. The basis of the immunoassay which can be performed with this redox label is a competition between analyte hapten and ferrocene-hapten conjugate for the antibodies and a signal generation proportional to the analyte concentration.

A model immunoassay was developed for antibodies against a ferrocene benzoic acid isothiocyanate conjugate (fer-benz, Ghindilis et al., 1995b). The conjugate is a good substrate for laccase and easily regenerated by quinoprotein glucose dehydrogenase; the lower detection limit is 0.5 nmol/l. After an external preincubation of fer-benz with antibody for 15 min the solution was transferred to the measuring cell. The response reflected the concentration of unbound conjugate, which is inversely proportional to the antibody concentration. The effect of antibody on the electrode response was examined without preincubation as well. In this case the antibody solution was injected directly into the batch and the response recorded for 10 nmol/l conjugate. Antibody without conjugate did not cause any response.

The results indicate that ferrocene derivatives can be used to label haptens, and the combination with a glucose dehydrogenase/laccase sensor is promising for the development of homogeneous immunoassays in the nanomolar concentration range.

Conclusion

Enzymatic substrate regeneration is a tool to enhance the sensitivity of enzyme electrodes both for substrate analysis and immunoassays.

Monoenzymatic cycles, where one enzyme catalyzes forward and backward reactions when the respective reagents are present (Schubert et al., 1990a, b) might facilitate the design of novel, simple enzymatic amplification schemes. On the other hand, the multiplication of the amplification by a second recycling enzyme pair in a double amplification system may yield a further increase in sensitivity. This has already been accomplished with pyruvate kinase/hexokinase-lactate dehydrogenase/lactate oxidase.

Further progress is expected from engineering appropriate recognition and signal generating systems, such as hybrid enzymes generated either by chemical 'site to site' linkage of interacting enzymes or by gene fusion. In this way the biochemical signal may be effectively channelled, which may lead to high sensitivity and specificity.

Acknowledgements

The authors thank A. Makower, A. Eremenko, A. Ghindilis, C. Bauer, F. Bier and E. Ehrentreich-Förster, who developed most of the described electrochemical immunoassays and contributed to catecholamine detection (also J. Szeponik and D. Pfeiffer) and B. Neumann for her work on the carbon paste electrodes. Financial support by German BMBF (0310821, 0319579 A, and 0310822) is gratefully acknowledged.

References

Athey, D. and McNeil, C. (1994) Amplified electrochemical immunoassay for thyrotropin using thermophilic β-NADH oxidase. *J. Immunol. Meth.* 176: 153–162.

Bauer, C.G., Eremenko, A.V., Ehrentreich-Förster, E., Bier, F.F. Makower, A., Halsall, H.B., Heineman, W.R. and Scheller, F.W. (1996) Zeptomole-detecting biosensor for alkaline phosphatase in an electrochemical immunoassay for 2,4-dichlorophenoxyacetic acid. *Anal. Chem.* 68: 2453–2458.

Bier, F. and Scheller, F. (1996) Bi-enzyme amplification cycles based on oligosaccharide dehydrogenase. *Fres. Z. Anal. Chem.* 354: 861–865.

Bier, F., Förster, E., Makower, A. and Scheller, F. (1996) An enzymatic amplification cycle for highly sensitive immunoassay. *Anal. Chim. Acta*, 328: 27–32.

Chen, F.C., Lin, N.N., Kuo, J.S., Cheng, L.J., Chang, F.M. and Chia, L.G. (1994) Simultaneous measurement of plasma serotonin, catecholamines, and their metabolites by an in vitro microdialysis-microbore HPLC and a dual amperometric detector. *Electroanal.* 6: 871–877.

Ciolkowski, E.L., Maness, K.M., Cahill, P.S., Wightman, R.M., Evans, D.H., Fosset, B. and Amatore, C. (1994) Disproportionation during electrooxidation of catecholamines at carbon-fiber microelectrodes. *Anal. Chem.* 66: 3611–3617.

Conrath, N., Gründig, B., Hüwel, S. and Cammann, K. (1995) A novel enzyme sensor for the determination of inorganic phosphate. *Anal. Chim. Acta* 309: 47–52.

Dennison, M.J., Hall, J.M. and Turner, A.P.F. (1995) Gas-phase microbiosensor for monitoring phenol vapor at ppb levels. *Anal. Chem.* 67: 3922–3927.

Eremenko, A.F., Makower, A., Jin, W., Rüger, P. and Scheller, F.W. (1995 a) Biosensor based on an enzyme modified electrode for highly-sensitive measurement of polyphenols. *Biosens. Bioelectr.* 10: 717–722.

Eremenko, A.F., Makower, A. and Scheller, F.W. (1995 b) Measurement of nanomolar diphenols by substrate recycling coupled to a pH-sensitive electrode. *Fresenius J. Anal. Chem.* 351: 729–731.

Eremenko, A.F., Makower, A., Bauer, C., Kurochkin and Scheller, F.W. (1996) A bienzyme electrode for tyrosine containing peptides determination. *Electroanal.* submitted.

Ewing, A.G., Strein, T.G. and Lau, Y.Y. (1992) Analytical chemistry in microenvironments: Single nerve cells. *Acc. Chem. Res.* 25: 440–447.

Ghindilis, A.L., Makower, A. and Scheller, F. (1995 a) Nanomolar determination of ferrocene derivatives using a recycling enzyme electrode. Development of a redox label immunoassay. *Anal. Lett.* 28: 1–11.

Ghindilis, A.L., Makower, A., Bauer, C.G., Bier, F.F. and Scheller, F. (1995 b) Picomolar determination of p-aminophenol and catecholamines based on recycling enzyme amplification. *Anal. Chim. Acta* 304: 25–31.

Ghindilis, A.L., Makower, A. and Scheller, F. (1995 c) A laccase-glucose dehydrogenase recycling-enzyme electrode based on potentiometric mediatorless electrocatalytic detection. *Anal. Meth. Instr.* 2: 129–132.

Gorton, L. (1995) Carbon paste electrodes modified with enzymes, tissues, and cells. *Electroanal.* 7: 23–45.

Hall, G.F., Best, D.A. and Turner, A.P.F. (1988) Amperometric enzyme electrode for the determination of phenols in chloroform. *Enzyme Microb. Technol.* 10: 543–546.

Hopkins, T.R. (1985) A multipurpose enzyme sensor based on alcohol oxidase. *Int. Biotech. Lab.* 3: 20–25.

Ikeda, T., Katasho, I., Kamei, M. and Senda, M. (1984) Electrocatalysis with glucose oxidase immobilized graphite electrode. *J. Electroanal. Chem.* 48: 1969–1979.

Ikeda, T., Matsushita, F. and Senda, M. (1991) Amperometric fructose sensor based on direct bioelectrocatalysis. *Biosens. Bioelectr.* 6: 299–304.

Jankowski, J.A., Tracht, S. and Sweedler, J.V. (1995) Assaying single cells with capillary electrophoresis. *TRAC* 14: 170–176.

Jin, W., Bier, F., Wollenberger, U. and Scheller, F. (1995) Construction and characterization of a multi-layer enzyme electrode: Covalent binding of quinoprotein glucose dehydrogenase onto gold electrodes. *Biosens. Bioelectr.* 10: 823–829.

Kaisheva, A., Iliev, I., Kazareva, R., Christov, S., Petkova, J., Wollenberger, U. and Scheller, F. (1996) Enzyme/gas-diffusion electrodes for determination of phenol. *Sensor. Actuator.*, in press.

Kopp, L.E. and Miech, R.P. (1972) Nonlinear enzymatic cycling systems: the exponential cycling system. *J. Biol. Chem.* 247:3558–3563.
Kulys, J.J., Sorochinskii, V.V. and Vidziunaite, R.A. (1986) Transient response of bienzyme electrodes. *Biosensors* 2:135–146.
Liu, A.H. and Wang, E.K. (1994) Amperometric detection of catecholamines with liquid chromatography at a polypyrrole-phosphomolybdic anion-modified electrode. *Anal. Chim. Acta* 296:171–180.
Lowry, O.H. and Passonneau, J.V. (1972) *A Flexible System of Enzymatic Analysis*. Academic Press, New York.
Macholan, L. (1990) Phenol-sensitive enzyme electrode with substrate recycling for quantification of certain inhibitory aromatic acids and thio compounds. *Coll. Czech. Chem. Commun.* 55:2152–2159.
Makower, A., Bauer, C.G., Ghindilis, A. and Scheller, F. (1994) Sensitive detection of cocaine by combining conjugate displacement and an enzyme substrate recycling sensor. *In: Biosensors 94, Proceedings of the 3rd World Congress on Biosensors.* Elsevier Advanced Technology, Oxford.
Makower, A., Eremenko, A.V., Streffer, K., Wollenberger, U. and Scheller, F. (1995) Tyrosinase-glucose dehydrogenase substrate recycling biosensor. Highly sensitive measurement of phenolic compounds. *J. Chem. Tech. Biotech.* 65:39–44.
Matysik, F.M., Nagy, G. and Pungor, E. (1992) Analytical distinction between different catechols by means of reverse differential-pulse voltammetry. *Anal. Chim. Acta* 264: 177–184.
Mizutani, F., Yamanaka, T., Tanabe, Y. and Tsuda, K. (1985) An enzyme electrode for L-lactate with a chemically amplified response. *Anal. Chim. Acta* 177:153–166.
Mizutani, F., Yabuki, S. and Asai, M. (1991) Highly-sensitive measurement of hydroquinone with an enzyme electrode. *Biosens. Bioelectr.* 6:305–310.
Mizutani, F., Yabuki, S. and Katsura, T. (1993) Amperometric enzyme electrode with the use of dehydrogenase and NAD(P)H oxidase. *Sensor. Actuator. B* 13–14:574–575.
Moore, T.J., Nam, G.G., Pipes and Coury Jr., L.A. (1994) Chemically amplified voltammetric enzyme electrodes for oxidizable pharmaceuticals. *Anal. Chem.* 66:3158–3163.
Niwa, O., Morita, M. and Tabei, H. (1991) Highly sensitive and selective voltammetric detection of dopamine vertically separated interdigitated array electrodes. *Electroanal.* 3:163–168.
Önnerfjord, P., Emneus, J., Marko-Varga, G. and Gorton, L. (1995) Tyrosinase graphite-epoxy based composite electrodes for detection of phenols. *Biosens. Bioelectr.* 10:607–619.
Ortega, F., Dominguez, E., Jönsson-Pettersson, G. and Gorton, L. (1993) Amperometric biosensor for the determination of phenolic compounds using a tyrosinase graphite electrode in a flow injection system. *J. Biotech.* 31:289–300.
Ortega, F., Dominguez, E., Burestedt, E., Emneus, J., Gorton, L. and Marko-Varga, G. (1994) Phenol oxidase-based biosensors as selective detection units in column liquid chromatography for the determination of phenolic compounds. *J. Chromatogr.* 675:65–78.
Pfeiffer, D., Scheller, F., McNeil, C. and Schulmeister, T. (1994) Cascade like exponential substrate amplification in enzyme electrodes. *Biosens. Bioelectr.* 10:169–180.
Pfeiffer, D., Scheller, F.W., Wollenberger, U., Makower, A., Bier, F.F., Szeponik, J., Klimes, N., Gajovic, N., Ghindilis, A. and Scholz, C. (1996) Enzyme sensors for the detection of subnano- and milimolar concentrations. Development and application of biosensors. *Analyst*, in press.
Raba, J. and Mottola, H.A. (1994) On-line enzymatic amplification by substrate cycling in a dual bioreactor with rotation and amperometric detection. *Anal. Biochem.* 220:297–302.
Rosen, I. and Rishpon, J. (1989) Alkaline phosphatase as a label for a heterogeneous immunoelectrochemical sensor. *J. Electroanal. Chem.* 258:27–39.
Ruzgas, T., Emneus, J., Gorton, L. and Marko-Varga, G. (1995) The development of a peroxidase biosensor for monitoring phenol and related aromatic compounds. *Anal. Chim. Acta* 311:245–253.
Scheller, F. and Schubert, F. (1989) *Biosensoren*. Akademie Verlag. Berlin.
Scheller, F., Wollenberger, U., Schubert, F., Pfeiffer, D. and Bogdanovskaya, V.A. (1987) Amplification and switching by enzymes in biosensors. *GBF Monographs* 10:39–49.
Scheller, F., Schubert, F., Weigelt, D., Mohr, P. and Wollenberger, U. (1988) Molecular recognition and signal processing in biosensors. *Makromolek. Chem.* 17:429–439.

Scheller, F., Pfeiffer, D., Hintsche, R., Dransfeld, I. and Wollenberger, U. (1990) Analytical aspects of internal signal processing in biosensors. *Ann. NY Acad. Sci.* 613:68–78.

Scheller, F.W., Schubert, F., Pfeiffer, D., Wollenberger, U., Renneberg, R., Hintsche, R. and Kühn, M. (1992) Fifteen years of biosensor research in Berlin-Buch. *GBF Monographs* 17:3–10.

Scheller, F.W., Makower, A., Ghindilis, A.L., Bier, F.F., Förster, E., Wollenberger, U., Bauer, C.G., Micheel, B. Pfeiffer, D., Szeponik, J., Michael, N. and Kaden, H. (1995) Enzyme sensors for subnanomolar concentrations. *In*: *ACS Symposium Series* 613, Chapter 7, pp 70–81.

Schubert, F., Kirstein, D., Schröder, K.L. and Scheller, F. (1985) Enzyme electrodes with substrate and coenzyme amplification. *Anal. Chim. Acta* 169:391–396.

Schubert, F., Kirstein, D., Scheller, F., Appelqvist, R., Gorton, L. and Johansson, G. (1986) Enzyme electrodes for L-glutamate using chemical redox mediators and enzymatic substrate amplification. *Anal. Lett.* 19:1273–1288.

Schubert, F., Wollenberger, U., Scheller, F. and Müller, H.G. (1990a) Artificially coupled enzyme reactions with immobilized enzymes: biological analogs and technical consequences. *In*: D. Wise, (ed.) *Bioinstrumentation and Biosensors*. Marcel Decker, New York.

Schubert, F., Scheller, F. and Krasteva, N. (1990b) Lactate-dehydrogenase-based biosensors for glyoxylate and NADH determination: A novel principle of analyte recycling. *Electroanal.* 2:347–351.

Skladal, P. (1991) Mushroom tyrosinase-modified carbon paste electrode as amperometric biosensor for phenols. *Coll. Czech. Chem. Comm.* 56:1427–1433.

Tang, H.T., Lunte, C.F., Halsall, H.B. and Heineman, W.R. (1988) p-Aminophenyl phosphate: An improved substrate for electrochemical immunoassays. *Anal. Chim. Acta* 214: 187–195.

Tang, X. and Johansson, G. (1995) Enzyme electrode for amplification of NAD^+/NADH using glycerol dehydrogenase and diaphorase with amperometric detection. *Anal. Lett.* 28:2595–2606.

Uchiyama, S., Hasebe, Y., Shimizu, H. and Ishihara, H. (1993) Enzyme-based catechol sensorbased on the cyclic reaction between catechol and 1,2-benzoquinone, using L-ascorbate and tyrosinase. *Anal. Chim. Acta* 276:341–345.

Vidziunaite, R.A. and Kulys, J.J. (1985) Kinetic regularities of cyclic substrate conversion in enzyme membranes (russ). *Liet. TSR Mokslu Akad. darbai Ser.* C2:84–91.

Wang, J. and Chen. Q. (1995a) Highly sensitive biosensing of phenolic compounds using bioaccumulation/chronoamperometry at a tyrosinase electrode. *Electroanal.* 7:746–749.

Wang, J. and Chen, Q. (1995b) Micofabricated phenol biosensors based on screen printing of tyrosinase containing carbon ink. *Anal. Lett.* 28:1131–1142.

Wang, J., Lin, Y. and Chen, Q. (1993a) Organic-phase biosensors based on entrapment of enzymes within poly(ester-sulfonic acid) coatings. *Electroanal.* 5:23–28.

Wang, J., Lin, Y., Eremenko, A.V., Ghindilis, A.L. and Kurochkin, I.N. (1993b) A laccase electrode for organic-phase enzymatic assays. *Anal. Lett.* 26 : 197 – 207.

Wasa, T., Akimoto, K. Yao, T. and Murao, S. (1984) Development of laccase membrane electrode by using carbon electrode impregnated with epoxy resin and its response characteristics. *Nippon Kagaku Koishi* 9:1398–1403.

Wightman, R.M., Finnegan, J.M. and Pihel, K. (1995) Monitoring catecholamines at single cells. *TRAC* 14(4):154–158.

Wollenberger, U., Schubert, F., Scheller, F., Danielsson, B. and Mosbach, K. (1987a) Coupled reactions with immobilized enzymes in biosensors. *Studia biophys.* 119:167–170.

Wollenberger, U., Schubert, F., Scheller, F., Danielsson, B. and Mosbach, K. (1987b) A biosensor for ADP with internal substrate amplification. *Anal. Lett.* 20:657–668.

Wollenberger, U., Scheller, F., Pavlova, M., Müller, H.G., Risinger, L. and Gorton, L. (1989) Glutamate oxidase based biosensors. *GBF Monographs* 13:33–36.

Wollenberger, U., Schubert, F. and Scheller, F. (1992) Biosensor for sensitive phosphate detection. *Sensor. Actuator. B* 7:412–415.

Wollenberger, U., Schubert, F., Pfeiffer, D. and Scheller, F. (1993) Enhancing biosensor performance using multienzyme systems. *TIBtech* 11:255–262.

Wollenberger, U., Paeschke, M. and Hintsche, R. (1994) Interdigitated array electrodes for the determination of enzyme activites. *Analyst* 119:1245–1249.

Wollenberger, U., Neumann, B. and Scheller, F. (1995) Quinoprotein glucose dehydrogenase modified carbon paste electrodes. *Proc. Int. 6th Beijing Conference and Exhibition on Instrumental Analysis*, Beijing F 201–202.

Yamaguchi, S., Ozawa, S., Ikeda, T. and Senda, M. (1992) Sensitive amperometry of 4-aminophenol based on catalytic current involving enzymatic recycling with diaphorase and its application to alkaline phosphatase assay. *Anal. Sci.* 8:87–88.

Yang, X., Pfeiffer, D., Johansson, G. and Scheller, F.W. (1991) Enzyme electrodes for ADP/ATP with enhanced sensitivity due to chemical amplification and intermediate accumulation. *Electroanal.* 3:659–663.

Yaropolov, A.I., Kharybin, A.N., Emneus, J., Marko-Varga, G. and Gorton, L. (1995) Flow-injection analysis of phenols at a graphite electrode modified with co-immobilized laccase and tyrosinase. *Anal. Chim. Acta* 308:137–144.

Yao, T., Yamamoto, H. and Wasa, T. (1989) Chemical amplified enzyme electrodes by substrate recycling. *Proceedings ISE-Meeting, Kyoto*. Kodanasha, Tokyo, pp 1014–1015.

Frontiers in Biosensorics II
Practical Applications
ed. by. F.W. Scheller, F. Schubert and J. Fedrowitz

Thermistor-based biosensing

B. Danielsson and B. Xie

Pure and Applied Biochemistry, Lund University, S-22100 Lund, Sweden

Summary. In this review a universal thermistor-based biosensor system is described with examples from clinical chemistry, bioprocess monitoring and environmental control. The technique is based on the measurement of the small temperature changes associated with enzymatic reactions occurring in a microreactor with immobilized enzyme. The system has good operational stability and a sensitivity that permits measurements down to 1 μM concentrations. Current developments include devices constructed by micromachining for multisensing purposes and miniaturised instrumentation intended for use in portable monitoring. With use of special supports for enzyme immobilisation even untreated whole blood samples can be applied. Another current line of investigation involves hybrid biosensors, such as combinations of electrochemistry and calorimetry into bioelectrocalorimetric devices with interesting new properties.

Introduction

There is a rapidly growing need for sensors suitable for specific monitoring and control in biotechnology as well as in the biomedical field, for decentralised clinical chemistry and for *in vivo* monitoring of, in particular, glucose and lactate. In these fields heat sensitive biosensors, such as the enzyme thermistor (ET) have definite advantages. Since biological processes are generally more or less exothermic, the combination of a specific biocatalyst with a heat or temperature sensor leads to a biosensor with general applicability that can be used for the determination of a particular compound present in a complex mixture, irrespective of the optical and electrochemical properties of the sample. In addition, the temperature transducer is virtually free from drift and fouling since it does not need to be directly immersed in the sample or any other fluid. Of the many different approaches to simplified calorimetric devices for use with immobilised enzymes that have been tested since the 1970s (Mosbach and Danielsson, 1981), especially the so-called enzyme thermistor (Danielsson, 1990) has continued to develop. However, more widespread use or commercialisation of thermistor-based sensors has to date not been realised in spite of their attractive features, a situation that may change when the recent trend of miniaturisation of calorimetric devices has become fully developed (Xie et al., 1993 a).

Instrumentation

The thermistor-based calorimeter used in most of the studies described herein has previously been described by Danielsson (1990). It consists of a thermostated thickwalled aluminium jacket, 80 mm in diameter and 250 mm long, containing a cylindrical aluminium heat sink with heat exchangers and two column positions (Fig. 1). The two columns can be used either with different enzymes for two different assays or with one column acting as a reference column (split-flow). The columns are attached at the end of the thermistor probes and are readily exchangeable. The thermistors are connected to a Wheatstone bridge with a maximum sensitivity of 100 mV/m°C. Commonly used full-scale sensitivities are in the 10–50 m°C range permitting determinations in the 0.01–100 mM range. A large excess of

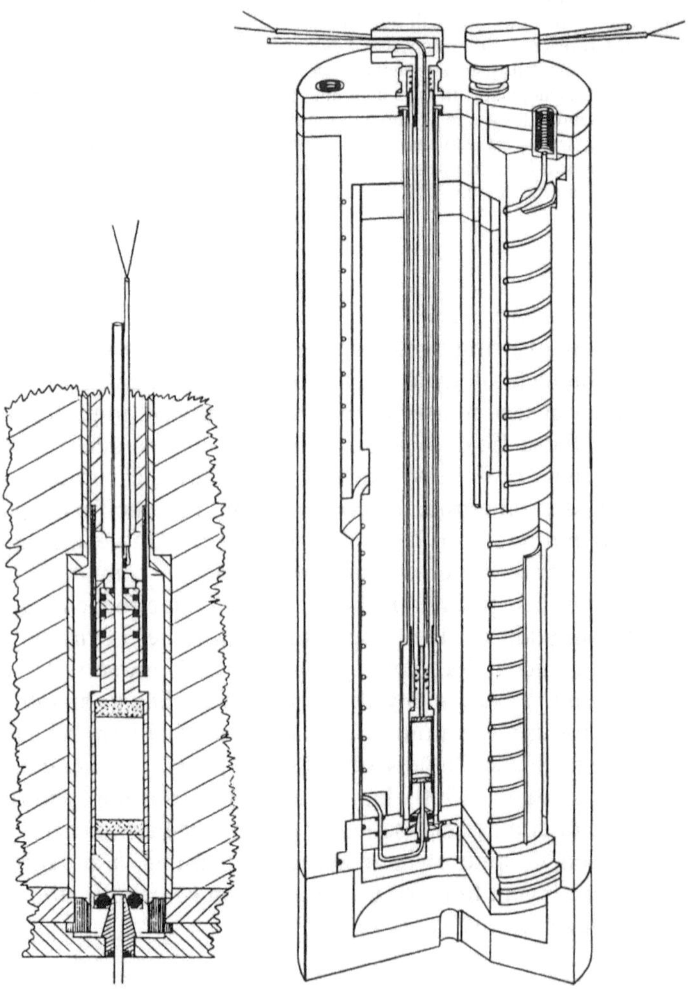

Figure 1. Cross-section of the calorimeter of a thermistor-based biosensor system with an aluminium constant temperature jacket and aluminium heat sink. The enlargement shows the attachment of a column and the transducer arrangement (the thermistor is fixed to the outlet gold tube with heat-conducting epoxy).

enzyme (10–100 units or more per column, column volume up to 1 ml) bound to a mechanically stable, highly porous support, such as controlled pore glass (CPG) or Eupergit C is used to insure an excellent operational stability. A suitable flow-for-flow injection analysis is 0.5–2 with a sample volume of 0.1–1 ml or smaller. Such small sample volumes will not lead to a thermal steady state, but result in a temperature peak that is proportional to the substrate concentration. A number of instruments of this type have been built at our institute and sold on a non-profit basis to various laboratories around the world. Efforts to lift up the sale to a more commercial level have not yet been successful.

Determination of metabolites

A large number of thermistor-based biosensor assays using immobilised enzyme reactors have been proposed for use in biotechnology, clinical chemistry and food analysis (Table 1 gives some examples). Oxidases generally offer higher sensitivity than dehydrogenases due to larger reaction heat (–ΔH 75–100 kJ/mol) and have no extra co-factor requirement. Co-immobilisation of oxidases with catalase has three additive effects: doubled total reaction heat by adding the enthalpy change of the catalase reaction (–100 kJ/mol); elimination of the hydrogen peroxide formed in the oxidase reaction avoiding protein damage by the hydrogen peroxide; and 50% improvement of the use of the oxygen available with a corresponding extention of the linear range. The low solubility of oxygen in aqueous solutions is, however, a serious drawback for the use of oxidases. Hydrolytic enzymes, such as disaccharidases, are usually associated with low enthalpy changes and have to be supplemented with secondary enzymes for practically useful assays.

A common way to increase the sensitivity of calorimetric measurements if a proteolytic reaction is involved, is to use buffers with high protonation enthalpy, such as Tris buffer, which has –47.5 kJ/mol as compared to –4.7 kJ/mol for phosphate buffer (Danielsson, 1995). Substrate and coenzyme recycling is another way of increasing the sensitivity, in favorable cases up to several 1000-fold. As an example 5000-fold amplification was observed by Scheller et al. (1985) using co-immobilised lactate oxidase (that oxidises lactate to pyruvate), lactate dehydrogenase (that reduces pyruvate to lactate) and catalase. Lactate (or pyruvate) concentrations as low as 10 nM could be determined with this arrangement. Similar sensitivities for ATP (alternatively ADP) were obtained by Kirstein et al. (1989) with the enzyme couple pyruvate kinase and hexokinase. Highly sensitive detection of ATP/ADP, the same as with bioluminescence, was accomplished by coupling the two cycles mentioned so that the pyruvate formed in the pyruvate kinase/hexokinase was recycled in the LDH/LOD cycle. The practicability of recycling is unfortunately somewhat limited since it is

Table 1. Linear concentration ranges of substances measured with enthalpimetric sensors using immobilized enzymes.

Analyte	Enzyme(s) used	Linear range (mM)
Ascorbic acid	Ascorbate oxidase	0.01–0.6
ATP (or ADP)	Pyruvate kinase + hexokinase	10 nM –*
Cellobiose	β-Glucosidase + glucose oxidase/catalase	0.05–5
Cephalosporins	Cephalosporinase (β-lactamase)	0.005–10
Creatinine	Creatinine iminohydrolase	0.01–10
Ethanol	Alcohol oxidase	0.002–1
Glucose	Hexokinase	0.01–25
Glucose	Glucose oxidase/catalase	0.001–1 (75**)
L-lactate	Lactate-2-mono-oxygenase	0.005–2
L-lactate	Lactate oxidase/catalase	0.002–1
L-lactate (or pyruvate)	Lactate oxidase/catalase + lactate dehydrogenase	10 nM–*
Oxalate	Oxalate oxidase	0.005–0.5
Penicillin	β-Lactamase	0.005–200
	Penicillin acylase	0.02–200
Sucrose	Invertase	0.05–100
Urea	Urease	0.005–200

* With substrate recycling; ** With benzoquinone as electron acceptor.

directly influenced by the actual activity of all enzymes involved and requires more frequent calibration than direct assays using immobilised enzyme reactors, where the sensitivity is virtually unchanged as long as there is excess enzyme activity. As discussed below, however, the enzymes involved in the LDH/LOD system are stable enough to make it practically useful, for instance as detecting mechanism in enzyme immunoassays.

Common enzyme supports and immobilisation procedures were briefly discussed by Danielsson (1991). More recent work has adopted a spherical CPG from Schuller GmbH (Steinach, Germany) with a particle size in the range of 125–140 μm and a pore size of 50 nm. This beaded support material produces enzyme columns with remarkable resistance to clogging by particles in the sample and permits even whole blood samples to be used (Xie et al., 1993b). The major limiting factor for column life is usually mechanical obstruction. If, however, the solutions used, as well as the samples, are filtered through a 1–5 μ or finer filter, and if microbial growth in the solutions and the flow lines is prevented, good operational stability with unchanged performance for large series of samples (thousands) can be obtained and the column can be used for several months.

In the following section some specific analysis are discussed. More details can be found in previous revies by Danielsson (1990 and 1991).

Glucose is the most common analyte in biosensor analysis and has also been subject to measurement by thermistor-based biosensors in various bioanalytical applications. It is usually carried out with glucose oxidase coimmobilised with catalase which provides high sensitivity (1 μM) and specificity. There is no cofactor requirement and the enzyme columns are very stable. Alternatively, the enzyme hexokinase can be used which, however, requires the cofactor ATP, but on the other hand a linear range of up to 25 mM can be obtained under usual conditions in contrast to about 1 mM with glucose oxidase/catalase. Hexokinase can also be used in an indirect assay for ATP with micromolar sensitivity, if the sample solution is supplemented with an excess of glucose. NADH can be measured in an analogous way with the same sensitivity using a lactate dehydrogenase column and excess of pyruvate.

The heat produced by the hydrolysis of cellobiose with β-glucosidase is too low to give sufficient sensitivity in a direct assay. By measuring the glucose formed in a precolumn containing β-glucosidase with a glucose oxidase/catalase loaded enzyme thermistor a typical operating range of about 0.05–5 mM can be obtained. The situation is the same with most other disaccharide splitting enzymes. Invertase (EC 3.2.1.26) produces, however, enough heat to allow direct determinations of sucrose in the range of 0.05 to 100 mM in an assay that it is not disturbed by the presence of glucose. Invertase columns are extremely stable and useful in food and bioprocess analysis.

L-Lactate can be determined with a sensitivity down to micromolar concentrations with two different enzyme systems: lactate-2-mono-oxygenase (EC 1.13.12.4) from *Mycobacterium smegmatis* and lactate oxidase from *Pediococcus pseudomonas* (EC 1.1.3.2) together with catalase. The latter enzyme is currently preferred because of its lower price and better availability, but the monooxygenase could be interesting for removal and simultaneous determination of lactate in combination with the previously described recycling arrangement for lactate/pyruvate. Since the product of the monooxygenase reaction is acetate and not pyruvate, this reaction could be used to remove the lactate from the sample.

Alcohols have been measured with alcohol oxidase (EC 1.1.3.13) from *Candida boidinii* or *Pichia pastoris*. Co-immobilisation with catalase increases the stability of the enzyme column to several months as discussed by Guilbault et al. (1983) who found a linear range of 0.005–1 mM (0.5 ml samples) using 0.1 M sodium phosphate, pH 7.0 as buffer. This allows for practical determinations of ethanol in beverages and blood and for the monitoring of fermentations.

Free cholesterol and cholesterol esters can be measured in detergent systems or in organic media. Cholesterol has been determined in phosphate buffer, pH 6.5, containing 12% (v/v) ethanol and 8% (v/v) Triton X-100 using cholesterol oxidase (EC 1.1.3.6) from *Nocardia erythropolis* or recently with an enzyme from *Streptomyces cinnamonensis*. Cholesterol esters can be- measured after hydrolysis with cholesterol esterase (EC 3.1.1.13). The measuring ranges are adequate for clinical use and it is possible to inject a serum sample directly into the aqueous buffer without detergent. Calibration has to be made with serum samples with known cholesterol content. Phospholipids and triglycerides can also be determined in clinical samples with use of thermometric biosensing.

Jack bean urease gives a linear range of at least 0.01–200 mM and offers a clinically useful assay that is independent of the ammonium concentration in the sample. Urease is very sensitive to inhibition by heavy metals, which has been turned into an advantage in a reversible procedure for heavy metal determination. On the other hand, the urease can be protected by addition of 1 mM EDTA and 1 mM reduced glutathione to the buffer leading to a very stable enzyme column. Acid urease from *Lactobacillus fermentum* has lower pH-optimum and somewhat different properties than Jack bean urease and has been used, especially in Japan, to remove urea from alkoholic beverages. This is important since urea (that may be present in beverages produced by fermentation) and ethanol upon standing or heating forms ethylcarbamate, a carcinogenic compound. Acid urease has been shown suitable for urea determination in such samples.

The thermometric assays of β-lactams (for instance penicillin G and V) using β-lactamases, such as penicillinase type I from *Bacillus cereus* (EC 3.5.2.6) have been particularly successful (Decristoforo and Danielsson, 1984). The useful linear range is about 0.005–200 mM. Several industrial applications have been developed using both discrete samples and continuous monitoring on pilot-plant and production scale fermentors. Although less exothermic, penicillin amidase (EC 3.5.1.11) is more specific and the alternative preferred by Rank et al. (1993) for measurements in fermentation broths. With both enzymes the columns are very stable and can be used for several months or for thousands of micro-filtered samples.

Several other metabolites have been measured with thermistor-based biosensors including ascorbic acid, cephalosporins, creatinine, galactose, hydrogen peroxide, lactose, malate, oxalate, uric acid, xanthine, and hypoxanthine.

Recycling-amplified TELISA

In the competitive TELISA (Thermometric Enzyme-Linked ImmunoSorbent Assay) method the ET column contains an immunosorbent. The sample is mixed with enzyme-labelled antigen and the concentration of bound antigen is determined by introduction of a substrate pulse, whereafter the column is regenerated by a pulse of glycine at low pH. The whole cycle takes only 13 minutes or less in the arrangement described by Birnbaum and co-workers (1986). The sensitivity is adequate for at-line determination of hormones, antibodies and other biomolecules produced by fermentation.

The use of alkaline phosphatase as an enzyme-label allow enhancement of the sensitivity by using phosphoenolpyruvate as substrate and the utiliza-

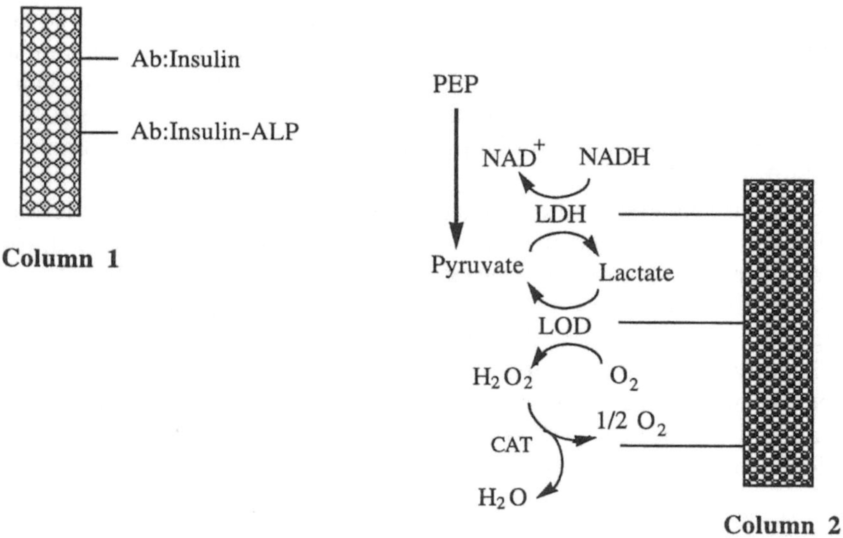

Figure 2. Principle scheme for a TELISA with enzymatic substrate recycling detection. Column 1 contains anti-insulin antibody (Ab) bound to Sepharose. Column 2 contains the three enzymes lactate dehydrogenase (LDH), lactate oxidase (LOD) and catalase (CAT) coimmobilized on controlled pore glass. PEP = phosphoenolpyruvate.

tion of a separate detection column in the ET-unit for the determination of the product (pyruvate) by substrate recycling. This is accomplished by using the substrate recycling system described above consisting of co-immobilized lactate dehydrogenase (reduces pyruvate to lactate under the consumption of NADH), lactate oxidase (oxidizes lactate to pyruvate), and catalase (Fig. 2). In addition, genetically engineered enzyme conjugates have been used in immunoassays. Thus a human proinsulin-*E. coli* alkaline phosphatase conjugate was used by Mecklenburg et al. (1993) for the determination of insulin or proinsulin. Concentrations lower than 1 µg/ml could be determined in less than 15 minutes.

On-line monitoring of bioprocesses

For on-line monitoring of bioprocesses using thermistor-based biosensors at a fermentation pilot-plant and at a production plant, the equipment was placed inside a steel cabinet flushed with cool, filtered air to keep the temperature sufficiently constant. The enzyme thermistor was automated and equipped with a pneumatic sampling valve and a sample selector. A sample stream of 0.5–2 ml/min was taken from an autoclavable 0.2 µ polypropylene hollow fiber filtration probe (Advanced Biotechnology Corp., Puchheim, Germany). In order to follow penicillin production, 0.1 ml samples were injected every 10–30 min over the duration of the fermentation (1–2 weeks). The flow through the ET unit (0.9 ml/min) was equally split between the enzyme column (β-lactamase or penicillin amidase bound to controlled pore glass) and an inactive reference column containing immobilized bovine serum albumin (Rank et al., 1993). The column was protected against microbial growth by adding 1 mM sodium azide to the buffer solution.

With this setup penicillin V could be measured during the entire fermentation run with the same enzyme column without serious problems in spite of rapid ambient temperature variations between 20 and 40 °C, high humidity and vibrations. The linear concentration range of penicillin V can be as large as 0.05–500 mM for 0.1 ml samples. The reference column efficiently compensates for nonspecific heat effects. Besides penicillin, measurements have been performed on *Saccharomyces* fermentations by Rank and co-workers (1993) using alcohol oxidase for ethanol, glucose oxidase for glucose and lactate oxidase for lactate. In all cases catalase was

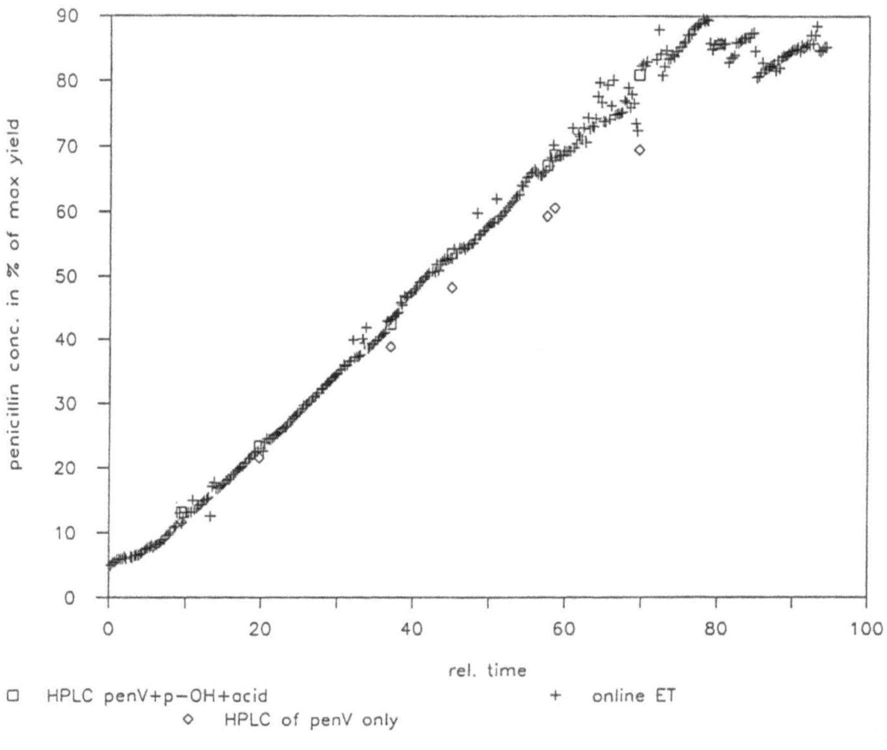

Figure 3. On-line monitoring of the penicillin V production in a 160-m³ bioreactor over 1–2 weeks using a thermistor-based biosensor equipped with a column containing penicillin V acylase. + = on-line biosensor values; = penicillin V, penicilloic acid and p-hydroxypenicillin measured off-line by HPLC; ◊ = off-line values of penicillin V measured by HPLC. The irregularities observed at the end of the run are due to discharge of fermentation broth and replacement with water (from Rank et al., 1992).

coimmobilized to increase sensitivity and linear range. In a recent study Rank et al. (1995) demonstrated on-line monitoring of ethanol, acetaldehyde and glycerol during industrial fermentations with *Saccharomyces cerevisiae*. Larger variations in the concentration registered could occassionally be seen, especially at the end of penicillin fermentations, when the viscosity of the broth was high due to very high cell mass, which may cause improper function of the filtration unit (Fig. 3). The general impression is, however, that the ABC filtration probe works better than a tangential flow unit at higher viscosities in smaller fermentors as well as in larger, production scale fermentors.

Characterization of an immobilised enzyme (invertase) was demonstrated in a study in which the kinetic constants were directly determined by a thermistor-based biosensor without the need for postcolumn analysis (Stefuca et al., 1990). In an extension of this work direct determination of the catalytic activity of immobilised cells was carried out (Gemeiner et al., 1993). *Trigonopsis variabilis* strains selected by mutagenesis for high cephalosporin transforming activity were used in a model system. The thermometric signal arising from the activity in one specific dominating enzymic step in the cells could be identified by comparative HPLC-analysis of the reaction mixture. The cephalosporin transforming activity by D-amino acid oxidase of selected yeast strains was identified in the same way. The thermometric signal was proportional to the number of cells as well as to the amount of enzyme (DAAO) immobilised in the ET minicolumn.

Measurements in organic solvents

Measurements in organic media using biosensors have attracted much interest in recent years. Thermistor-based biosensors, such as the ET, are of special interest since the temperature response depends on the heat capacity of the system and the specific heat is up to 3 times lower in some organic solvents than in water. In addition, the solubility of some enzyme substrates (cholesterol and triglycerides for instance) is higher in organic solvents than in water. It has been possible to design potentially useful procedures for enzyme analysis in organic solvents, especially since the enzymes involved may become stabilized by the immobilisation. It could happen that the enzymic activity is lost after some time, but it is often possible to restore it fully by treatment with aqueous buffer. The enthalpy change is likely to be different in organic solvents or in solvent-water mixtures than in pure buffer which makes it difficult to predict the temperature response. In a comparison of the temperature responses obtained for tributyrin in a buffer-detergent system and in cyclohexane with lipoprotein lipase immobilised on Celite the response was about 2.5 times higher in the latter case (as would be expected from the actual specific heats)-and linear up to higher concentration. In other experiments, however, the increase in sensitivity was found to be much higher.

The usefulness of calorimetric sensors for work in different media was demonstrated by Stasinska et al. (1989) in a study on immobilised α-chymotrypsin which was used for hydrolysis of peptide bonds in 0.05 M Tris-HCl, pH 7.8 containing 10% DMF and for syntheses of peptide bonds in 50% DMF + 50% 0.1 M Na-borate, pH 10.0. With the α-chymotrypsin immobilised in the enzyme thermistor column both reactions could be followed, the hydrolysis giving an exothermic response while the synthetic route was endothermic.

Environmental control

The effect of pollutants on biological reactions are clearly measureable (Danielsson and Mosbach, 1988), but only few routine applications of calorimetry has been described to date. One reason for this is the lack of suitable instrumentation. The inhibitory effect of pollutants such as pesticides on biological systems is usually irreversible which means that the column with immobilised enzyme or cells must be replaced after one positive sample resulting in an analysis speed of maybe only 1–2 samples per hour. Instruments with a magasine of columns or with several parallel columns could overcome drawbacks.

Metal detection with thermistor-based biosensors have mostly been performed in two ways: by measuring the inhibition of enzymic activity or by measuring the activation of apoenzymes by metal ions. In both cases a certain specificity can be obtained. An example of the first alternative is given by the very efficient inhibition of urease activity by heavy metal ions (Hg^{2+}, Cu^{2+} and Ag^{+}). By using enzyme columns with comparatively low activity, very sensitive determinations in the ppb range or lower can be made. The original activity of the urease column can be restored by washing with iodide and EDTA. The sampling frequency is 3–4 samples per hour (Danielsson and Mosbach, 1988).

Many enzymes require a certain metal in their active site to be active. It is often possible to remove this metal with strong chelating agents, which results in an inactive apoenzyme. Upon exposure to a sample containing the same (or related) metal ion, the activity is restored to an extent that is related to the concentration of the metal ion. This procedure can be repeated up to a couple of times per hour. Examples of this technique including determination of Co^{2+}, Cu^{2+} and Zn^{2+} at nanomolar concentrations have been given by Satoh (1989).

Miniaturized thermistor-based biosensors

During the last few years miniaturisation of biosensors has been accelerated primarily due to development in micromachining and semiconductor technologies. In this course, miniaturised thermal biosensors have shown strong potential and great prospective in bioanalysis, particularly in clinical biochemical analysis, decentralised health-care and bioprocess control. Furthermore, the universal detection principle provides a simple, stable and reliable means for determination of multiple analytes using an integrated thermal biosensor array coupled with the respective enzymes.

As an intermediary step in the investigation of highly miniaturised, micromachined constructions, less miniaturised plastic/aluminium devices in the size of 50 mm in length and 15–25 mm in diameter or smaller have been designed (Fig. 4). Due to high sensitivity, small dimensions, modest buffer consumption, and good operational stability these devices have been found suitable for portable use, for instance for home monitoring of glucose in diabetes. To allow analysis directly on whole blood samples three different approaches have been tested with good results.

In the first approach the blood cells are removed by dialysis or filtration using small coaxial dialysis units constructed by attaching a 25-mm long 0.2-mm (i.d.) cuprophan hollow fibre inside a 0.5-mm PVC tubing. These units give about 5% yield of the glucose in the sample resulting in a linear range of up to 25 mM glucose with a 1.5×15 mm glucose oxidase/catalase column, which is adequate for diabetes monitoring. It is also possible to use a microdialysis probe (CMA/Microdialysis, Stockholm, Sweden) in which a thin needle (0.6 mm diameter) with a dialysis tubing (4–30 mm long) is inserted in a vein or under the skin. Low-molecular weight compounds are transported to the sample valve of the analytical device by a slow buffer stream (typically < 5 µl/min) resulting in a linear range for glucose of about 1–25 mM and was found to provide a reliable method for *ex vivo* monitoring of glucose by Amine et al. (1995).

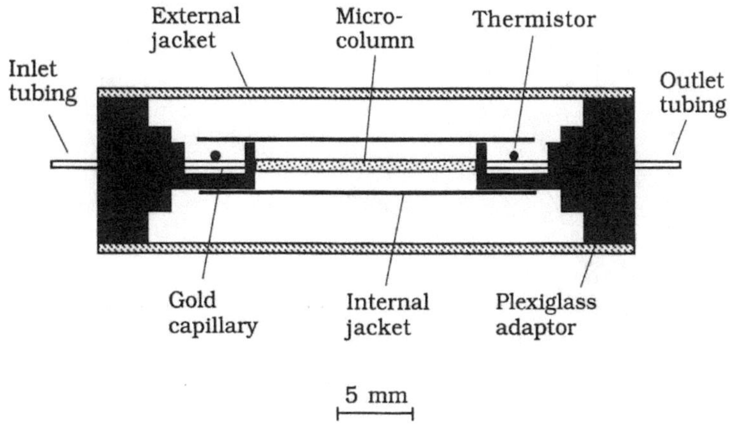

Figure 4. Schematic diagram of a miniaturized enzyme thermistor.

In a second line of study, a superporous agarose material developed at our department was investigated as enzyme carrier and was found to allow the injection of a large number of whole blood samples on the enzyme column without any sign of clogging. The calibration curve obtained with 20-μl samples injected in a flow of 100 μl/min was linear up to at least 25 mM glucose. Blood samples were applied 10-fold diluted.

In the third approach, a very small column (0.6 mm × 10 mm) was filled with spherical 125–175 micron CPG-particles loaded with glucose oxidase/catalase. Such a column provides spaces between the particles that are large enough to allow the cells to pass through without being trapped. As seen from Figure 5, a sample volume of 1 μl gave a suitable measuring range (1–25 mM) for blood glucose determination. Xie et al. (1993b) found that over 100 blood samples could be injected on this type of column. In conclusion all three approaches appear to be partically useful for a home monitoring device. Similar assays for urea (0.2–50 mM) and lactate (0.2–14 mM) using 1-μl samples have been developed by Xie et al. (1994).

More extensive miniaturisation of thermistor-based biosenor flow systems can be accomplished by micromachining in materials such as silicon and quartz as demonstrated by Xie and co-workers (1992). Recent developments include studies on multianalyte determination in a single sample stream by sequentially arranged enzyme systems (Fig. 6). Xie et al. (1995) described dual-analyte measurement examplifed by glucose and urea or penicillin and urea measurement with a linear range of up to 20 mM for urea, 8 mM for glucose and 40 mM for penicillin. Current designs permit simultaneous assay of as many as four analytes.

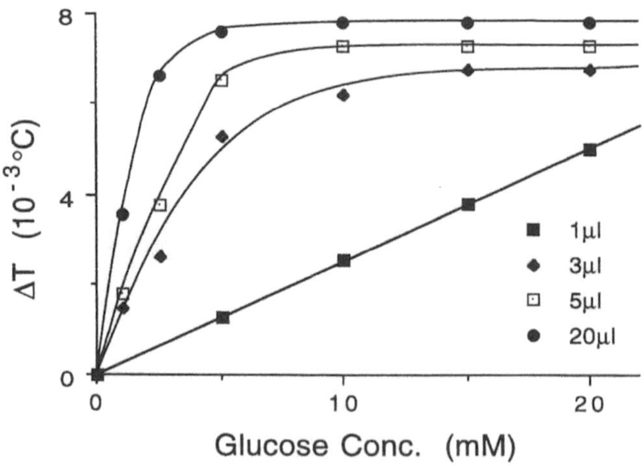

Figure 5.. Effect of sample volume on the linear range of a thermometric glucose sensor with a 0.6 mm × 10 mm CPG-column with glucose oxidase/catalase. The flow rate was 50 μl/min.

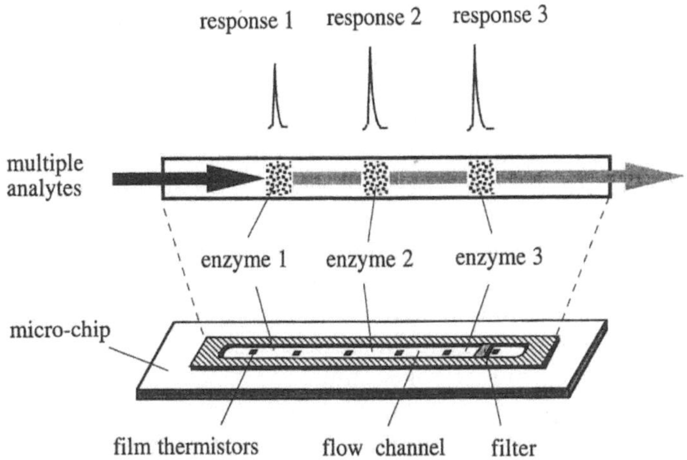

Figure 6. Working principle of a flow injection thermal microbiosensor for triple analyte determination. A pair of thermistors are placed upstream and downstream of each enzyme matrix. Enzyme-free regions separates the enzyme regions from each other.

Hybrid biosensors

Another field under current investigation is hybrid sensors which combine two different measurement technologies into a hybrid. Conventional biosenors are usualy classified into categories, such as electrochemical, optical, and thermal sensors according to the detection principle. Each type of sensor has its own merits and drawbacks. Creating hybrid biosenors by combination of different detection principles could possibly retain the original advantages and avoid the disadvantages of the respective type of sensor. Furthermore, hybridisation of biosenors could also create unique properties.

One example of a hybrid biosenor was recently fabricated and demonstrated by Xie and coworkers (1993 c). The biosensor combines electrochemical regeneration of the enzyme with flow injection biocalorimetry. The enzyme column that is constructed of electrically conductive materials functions as the working electrode, enzyme catalytic reactor and together with a thermistor as a thermally sensitive element. Reticulated vitreous carbon (RVC) was used as support onto which the enzymes and electron mediators were immobilized or adsorbed and packed into a platinum column. The column and a platinum counter electrode were connected to a simple potentiostat to carry out the electrochemical regeneration of the mediators. Simultaneously, the temperature changes in association with the enzyme reactions were differentially detected by a pair of thermistors which were mounted at the inlet and outlet of the column, respectively. This bioelectrocalorimetric device was tested for glucose determination using glucose oxidase with ferrocene as mediator. This sensor was less susceptible to interferences than a conventional electrochemical device and the linear range was independent of the oxygen concentration in contrast to a normal thermal biosenor which is limited by the oxygen concentration in buffer.

In a recent study a hybrid biosenor was constructed based on tyrosinase catalyzed catechol reactions. Catechol was electrochemically regenerated from 1,2-benzoquinone which was produced by oxidation of catechol by tyrosinase. The current and temperature changes in relation to the reactions were simultaneously detected. The results indicated that using electrocatalytic recycling the thermal detection could be improved in sensitivity and linear range. The thermal signal, on the other hand, could be used as a reference for the electrical signal, for example for the calculation of the recycling factor, provided there is no contribution to the temperature signal from the electrochemical catalysis. In our experiment we could not demonstrate any significant difference between the thermal signals obtained with and without applying the electrochemical catalysis.

Acknowledgement
The continuous support from the Swedish National Board for Technical Developments and all who have contributed to this research field are gratefully acknowledged.

References

Amine, A., Digua, K., Xie, B. and Danielsson, B. (1995) A microdialysis probe coupled with a miniaturized thermal glucose sensor for *in vivo* monitoring. *Anal. Lett.* 28: 2275–2286.

Birnbaum, S., Bülow, L., Hardy, K., Danielsson, B. and Mosbach, K. (1986) Rapid automated analysis of human proinsulin produced by *Escherichia coli*. *Anal. Biochem.* 158: 12–19.

Danielsson, B. (1990) Calorimetric biosenors. *J. Biotechnol.* 15:187–200.

Danielsson, B. (1991) Enzyme thermistor devices. *In*: L.J. Blum and P.R. Coulet (eds): *Biosensor Principles and Applications*. Marcel Dekker, Inc., New York, pp 83–105.

Danielsson, B. (1995) *Encyclopedia of Analytical Science*. Academic Press, London, pp 4597–4605.

Danielsson, B. and Mosbach, K. (1988) Enzyme thermistors. *Methods Enzymol.* 137:181–197.

Decristoforo, G. and Danielsson, B. (1984) Flow injection analysis with enzyme thermistor detector for automated detection of β-lactams. *Anal. Chem.* 56:263–268.

Gemeiner, P., Stefuca, V., Welwardová, A., Michalková, E., Welward, L., Kurillová, L. and Danielsson, B. (1993). Direct determination of the cephalosporing transforming activity of immobilized cells with use of an enzyme thermistor. 1. Verification of the mathematical model. *Enzyme Microb. Technol.* 15:50–56.

Guilbault, G.G., Danielsson, B., Mandenius, C.F. and Mosbach, K. (1983) Enzyme electrode and thermistor probes for determination of alcohols with alcohol oxidase. *Anal. Chem.* 55:1582–1585.

Kirstein, D., Danielsson, B., Scheller, F. and Mosbach, K. (1989) Highly Sensitive Enzyme Thermistor Determination of ADP and ATP by Multiple Recycling Enzyme Systems. *Biosensors* 4:231–239.

Mecklenburg, M., Lindbladh, C., Li, H., Mosbach, K. and Danielsson, B. (1993) Enzymatic amplification of a flow-injected thermometric enzyme-linked immunoassay for human insulin. *Anal. Biochem.* 212:388–393.

Mosbach, K. and Danielsson, B. (1981) Thermal bioanalyzers in flow streams. Enzyme thermistor devices. *Anal. Chem.* 53:83A.

Rank, M., Gram, J. and Danielsson, B. (1993) Industrial on-line monitoring of penicillin V, glucose and ethanol using a split-flow modified thermal biosenor. *Anal. Chim. Acta* 281: 521–526.

Rank, M., Gram, J., Stern-Nielsen, K. and Danielsson, B. (1995) On-line monitoring of ethanol, acetaldehyde and glycerol during industrial fermentations with *Saccharomyces cerevisiae. Appl. Microbiol Biotechnol.* 42:813–817.

Satoh, I. (1989) Continuous biosensing of heavy metal ions with use of immobilized enzyme reactors as recognition elements. *Materials Res. Soc. Int. Meeting on Advanced Materials*, Tokyo 14:45.

Scheller, F., Siegbahn, N., Danielsson, B. and Mosbach, K. (1985) High-sensitivity enzyme thermistor assay of L-lactate by substrate recycling. *Anal. Chem.* 57:1740–1743.

Stasinska, B., Danielsson, B. and Mosbach, K. (1989) The ue of biosenors in bioorganic synthesis: Peptide synthesis by immobilized α-chymotrypsin assessed with an enzyme thermistor. *Biotechnol. Techn.* 3:281–288.

Stefuca, V., Gemeiner, P., Kurillová, L., Danielsson, B. and Báles, V. (1990) Application of the enzyme thermistor to the direct estimation of intrinsic kinetics using the saccharose-immobilized invertase system. *Enzyme Microb. Technol.* 12:830–835.

Xie, B., Danielsson, B., Norberg, P., Winquist, F. and Lundström, I. (1992) Development of a thermal micro-biosensor fabricated on a silicon chip. *Sensor. Actuator. B* 6:127–130.

Xie, B., Danielsson, B. and Winquist, F. (1993a) Miniaturized thermal biosensors. *Sensor. Actuator. B* 15–16:443–447.

Xie, B., Hedberg (Harborn), U., Mecklenburg, M. and Danielsson, B. (1993b) Fast determination of whole blood glucose with a calorimetric micro-biosensor. *Sensor. Actuator. B* 15–16:141–144.

Xie, B., Khayyami, M., Nwosu, T., Larsson, P.-O. and Danielsson, B. (1993c) Ferrocene-mediated thermal biosensor. *Analyst* 118:845–848.

Xie, B., Harborn, U., Mecklenburg, M. and Danielsson, B. (1994) Urea and lactate determined in 1-μL whole blood with a miniaturized thermal biosensor. *Clin. Chem.* 40:2282–2287.

Xie, B., Mecklenburg, M., Danielsson, B., Öhman, O., Norlin, P. and Winquist, F. (1995) Urea and lactate determined in 1-μL whole blood with a miniaturized thermal biosensor. *Analyst* 120:155–160.

Frontiers in Biosensorics II
Practical Applications
ed. by. F.W. Scheller, F. Schubert and J. Fedrowitz

Electrochemical gas biosensors

I. Iliev and A. Kaisheva

Central Laboratory of Electrochemical Power Sources, Bulgarian Academy of Sciences, 1113 Sofia, Bulgaria

Summary. Gas biosensors for detection of vapors of some volatile compounds (SO_2, alcohol, formaldehyde, phenol) are reviewed. The enzymes sulphite oxidase, alcohol oxidase, alcohol dehydrogenase, formaldehyde dehydrogenase and polyphenol oxidase are used in electrochemical cells separated from the gas phase by a porous membrane, and in microbiosensors with "enzyme gel" deposited onto an interdigitated gold two-electrode system. A gas biosensor for the vapors of phenolic compounds vapors, comprising an enzyme/gas-diffusion electrode with tyrosinase enzyme is investigated. The transient amperometric signal and the calibration curves of this gas biosensor are studied in the presence of phenol, p-cresol and 4-chlorophenol vapors. It is shown that phenol vapor concentrations in the ppb range are detectable with this type of gas biosensor.

Introduction

Environmental pollution is one of the most substantial problems of present-day society. Often soil and surface waters around chemical plants are contaminated. A number of volatile chemical compounds with high toxicity are emitted into the atmosphere. In this connection research and development of methods and devices for the assessment of pollution levels in all types of waters and in the atmospheric air is of prime importance.

Nowadays there is a great need for gas sensors for an efficient and effective detection and on-line monitoring of hazardous gases in the ambient air. Monitoring of the volatile components evolved from food can be used for the characterization of food freshness. On the other hand the analysis of some volatile components in respiratory gas with a suitable gas sensor would provide a convenient and noninvasive method of diagnosis and monitoring of disease states.

A great variety of physical and chemical gas sensors have been developed: sensors based on colorimetric principles, semiconductor-type gas sensors, even artificial noses (Moseley et al., 1991; Vairavamurthi et al., 1992; Sigrist, 1994).

Electrochemical gas sensors exist for a limited range of electrochemically active gases (Malinski et al., 1992). However, in some cases these types of gas sensors may have low sensitivity and display a relative poor selectivity for analytes in complex gas mixtures. Therefore, there is a clear need for sensitive, selective and reliable gas-sensor in environmental and food control as well as in medicine.

Electrochemical biosensors are widely employed for the analysis of various chemical substances (such as glucose, cholesterol, lactate, alcohols, phenols etc.) in aqueous solutions in environmental monitoring, medicine and biotechnology (Scheller and Schubert, 1989). A great advantage of electrochemical biosensors based on biologically active recognition proteins, such as enzymes, is their high inherent specificity and selectivity in the analysis of their substrates. Recently some achievements have been reported in the field of electrochemical gas biosensors for the detection and monitoring of gaseous substances in the air or in respiratory gas.

Electrochemical gas biosensors for SO_2, ethanol and formaldehyde

Electrochemical gas biosensors based on sensors for SO_2, ethanol and formaldehyde interdigitated gold microelectrode systems have been developed (O'Sullivan et al., 1995) for the detection of SO_2, phenol, ethanol and formaldehyde. A two-electrode system is used in all cases consisting of one set of gold microbands acting as the working electrode and another set acting as a combined counter/quasi reference electrode. A gel layer containing both the enzyme immobilized in an organic/aqueous phase mixture and the electrolyte is applied to the working area of the screen-printed microelectrode system.

In our group such a type of gas biosensor using the enzyme sulfite oxidase and potassium hexacyanoferrate(III) as mediator was investigated for the determination of SO_2. The enzyme and the electrochemical mediator were contained in agarose/carboxymethylcellulose gel. A "gas rig" capable of delivering known concentrations of SO_2 at different relative humidity and temperatures was employed. A linear calibration curve was obtained at a working potential of ca -80 mV vs Ag/AgCl in the SO_2 concentration range 0.75–15 ppm (at a relative humidity of 90%). Gas biosensors with similar design were investigated for the analysis of ethanol vapor (using alcohol oxidase or alcohol dehydrogenase) and for the analysis of formaldehyde (based on the enzyme formaldehyde dehydrogenase).

Another type of gas biosensor for ethanol has recently been reported (Mitsubayashi et al., 1994) consisting of a two-compartment cell (for liquid and gas phases) separated by a porous PTFE diaphragm membrane. The enzyme electrode is constructed using a commercially available Clark-type dissolved oxygen electrode. The enzyme alcohol oxidase is immobilized in an acrylamide membrane placed onto the sensing area of the electrode. The tip of the enzyme electrode is immersed into the liquid compartment filled with phosphate buffer. The sensor is calibrated in the range from 0.358 to 1242 ppm ethanol vapor and shows high selectivity for ethanol in the presence of other gases. It gives only a negligible response to n-pentane, methylethylketone, hexane and acetone. The observed short life-time of this gas sensor is due to the deactivation of the enzyme alcohol oxidase and its separation from the immobilization membrane.

Formaldehyde is a common industrial chemical. It is toxic and irritant to eyes and skin. A gas biosensor for the determination of formaldehyde in the atmosphere has recently been proposed (Hämmerle et al., 1996). The biosensor comprises a platinum black or graphite working electrode bonded to a porous Teflon membrane and combined with a graphite counter electrode and saturated calomel reference electrode. The enzyme formaldehyde dehydrogenase is immobilized on the working electrode surface by means of a dialysis membrane. The cell is designed for dry storage in order to ensure a long shelf-life of the biosensor. It is filled with electrolyte prior to use. The electrolyte contains 1 mM NAD^+, 1 mM 1,2-naphtoquinone-4-sulphonic acid as electrochemical mediator, 0.1 M KCl and 0.1 M potassium phosphate buffer. The working potential is 0.15 V vs SCE. The gaseous formaldehyde is sampled from the head space above an aqueous solution of known concentration. A 15-ml glass bottle with defined form is filled with 12 ml formaldehyde solution and sealed to the inlet of the gas biosensor. The formaldehyde concentration in the gas phase is calculated according to the equation given by Dong and Dasgupta (1986). It should be noted that the formaldehyde vapor concentration above its aqueous solution is temperature dependent, so that, in order to eliminate possible errors, the temperature must be thoroughly controlled during the measurements. The calibration curve of the formaldehyde gas biosensor is linear up to 6 vppm. The lower detection limit is about 0.3 vppm. At formaldehyde concentrations higher than 20 vppm the signal of the gas biosensor decreases, probably due to substrate inhibition. Reproducible results can only be obtained by using graphite working electrodes. If a platinum-black working electrode is used, the results are affected by the large current response of the Pt-electrode to methanol, which is present in the standard formaldehyde solutions as stabilizer.

Phenol vapor biosensors

Phenolic compounds are among the major pollutants in industrial waste from coke ovens, oil refineries, petrochemical production, plastic industry etc. Because of the environmental and toxicological significance of phenolic compounds strict limits of phenol concentration in waters and air are prescribed by law. In this connection the development of a highly selective and sensitive gas biosensor for phenol vapors, which may be based on the enzyme tyrosinase, is of great practical interest.

Recently a biosensor for measuring phenol vapor concentration was designed (Dennison et al., 1995) using the enzyme polyphenol oxidase. An interdigitated microelectrode array is used as transducer. It consists of a two-electrode system (interdigitated microelectrode) on which a small amount (4 µl) of an "enzyme gel" is applied, containing the enzyme polyphenol oxidase, phosphate buffer, 0.1 M KCl and glycerol. The measurements are performed at a working potential of –150 mV vs Ag/AgCl. A quite complicated "gas rig" for generation of phenol vapors at different relative humidities is used to obtain the calibration curve of the gas biosensor up to 14 ppm phenol vapor concentration. The conjunction of the enzymatic and an electrochemical reaction at the interdigitated microband electrodes gives the sensors an extremely high sensitivity. Measurements of the gas substrates in the ppb range can be performed.

It should be noted that the normal operation of the above-described gas biosensor is substantially affected by the relative humidity of the air. A decrease of the biosensor's amperometric signal is observed in all cases with a decrease in air humidity. At low relative humidity a rapid loss of water from the enzyme gel is observed, due to the continuous exchange of water molecules between the gel and the atmosphere. The change of the water content of the gel affects both the enzyme activity and the ion activity, which causes changes in the electrochemical potential of the reference/counter electrode. The observed disadvantage is inherent to open microelectrochemical systems which exchange water molecules with the surrounding air.

A new type of gas biosensor for determination of the concentration of vapors of some phenolic compounds in atmospheric air was recently designed by using an enzyme/gas-diffusion electrode with the enzyme tyrosinase (Kaisheva et al., 1995a; Illiev et al., 1995). The biosensor is a two-electrode electrochemical cell comprising a tyrosinase/gas-diffusion electrode and an Ag/AgCl electrode used as a counter and reference electrode; both electrodes are placed in an electrolyte space filled with 0.1 M phosphate buffer containing 0.1 M KCl. The enzyme/gas-diffusion electrode comprises two layers: a gas permeable porous gas layer made of a carbon material modified with PTFE by a special technology (Iliev et al., 1977) and a composite enzyme layer containing the enzyme tyrosinase, carbon material and Nafion 117 pressed on the supporting layer. The carbon gas layer possesses a very high porosity (0.95 cm^3/g) combined with a high hydrophobicity (wetting angle $\Theta = 116°$) and electronic conductivity. The pores of this carbon gas layer are hydrophobic and thus free of electrolyte, so that gases are rapidly transported through this layer to the zone where the enzyme is immobilized. The enzyme is immobilized strongly in the composite enzyme layer so that no losses of the enzyme are observed during the operation of the electrode. Mushroom tyrosinase (E.C.1.14.18.1) is used which catalyzes the oxidation of the phenolic compounds via hydroxylation with molecular oxygen to catechols and subsequent oxidation to the corresponding quinone compounds:

$$\text{Phenol} + O_2 \xrightarrow{\textit{tyrosinase}} \text{Catechol} \quad (1)$$

$$\text{Catechol} + O_2 \xrightarrow{\textit{tyrosinase}} o\text{-Quinone} \quad (2)$$

The electrochemical reducton of *o*-quinone produced in the above enzymatic reaction is used as a detection reaction by which an electrical signal proportional to the phenol concentration is obtained:

$$o\text{-Quinone} + H^+ + 2\,e^- \longrightarrow \text{Catechol} \qquad (3)$$

The catechol produced in the electrochemical reaction (3) also participates in the enzymatic reaction (2) so that an amplifying effect is observed (Dennison et al., 1995).

It should be noted that beside the phenol vapors also oxygen from atmospheric air is transported in the gas phase through the porous hydrophobic gas layer of the enzyme/gas-diffusion electrode to the enzyme layer. This fact is important since oxygen takes part in the enzymatic reaction (1) and (2). The supply of oxygen ensures a high oxygen concentration in the zone of the enzymatic reaction which allows a high of enzymatic oxidation of phenol to be achieved.

A similar tyrosinase/gas-diffusion electrode placed in a conic plastic tube has been used for the determination of phenol concentration in water solutions (Kaisheva et al., 1995a). In this case the enzyme/gas-diffusion electrode was mounted in the electrochemical cell in such a way that the enzyme layer (the narrow end of the conic tube) is in contact with the electrolyte. The opposite side of the conical tube from the side of the porous carbon gas layer is in contact with air. In this mode of operation the enzyme layer is supplied with the substrate, phenol, from the electrolyte while the oxygen needed for the enzymatic reaction is supplied to the immobilized enzyme (by a fast transport from the air) through the porous hydrophobic layer. The investigation of the tyrosinase/gas-diffusion electrode for determination of phenol concentration in water has shown that the rapid supply of gaseous oxygen to the zone of the enzymatic reaction results both in an increase of the amperometric signal of the phenol electrode and in an enlargement of the linear part of its calibration curve (Kaisheva et al., 1995a).

When used in a phenol gas biosensor the tyrosinase/gas-diffuson electrode is mounted in the electrochemical cell in such a way that the enzyme-containing layer contacts the electrolyte in the electrolyte space (ca 1 ml). The porous hydrophobic layer of the electrode is exposed to the surrounding gas. An amperometric signal is obtained when the biosensor is placed in the gas-space 3 cm above the level of the phenol solution (Fig. 1) in such a way that the hydrophobic gas layer of the enzyme/gas-diffusion electrode is facing the solution. In this case the signal of the biosensor is due to the presence of the phenol vapor above the solution. If the biosensor is placed above pure water, no amperometric signal is observed. The transients of the amperometric signal are for different concentrations at 25 °C. The experiments show the reproducibility of the amperometric signals for two different phenol concentrations (Fig. 2). The biosensor showed no response to a range of different solvent vapors including chloroform and acetone.

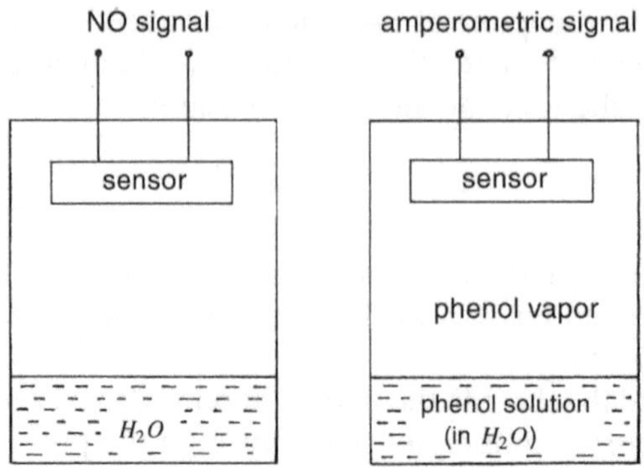

Figure 1. Schematic presentation of the test conditions of a phenol gas biosensor.

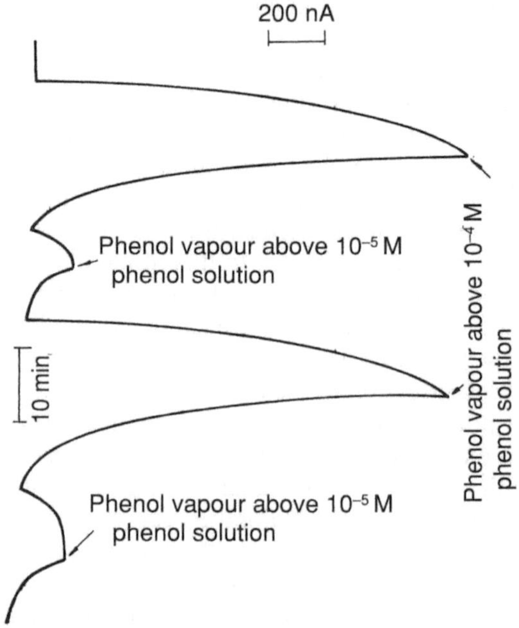

Figure 2. Transient response of a phenol gas biosensor when consecutively placed above 10^{-4} and 10^{-5} phenol solutions.

Polarization curves of the investigated gas biosensor measured when the sensor was placed above pure water (background curve) and above 10^{-4} phenol solution are presented in Figure 3. In the absence of phenol vapors and at positive potentials the current ($I_{background}$) is anodic and very low. At negative potentials the current is cathodic and increases substantially with the increase of the potential in cathodic direction due to the process of the electrochemical reduction of oxygen. At a potential 0.00 V vs Ag/AgCl the background current is practically zero. In the presence of phenol vapors a significant amperometric signal (I_{phenol}) is observed, due to the enzymatic oxidation of the phenol vapors. In the upper part of the same figure the net phenol signal $\Delta I = | I_{phenol} - I_{background} |$ is presented as a function of the potential. It can be seen the amperometric signal ΔI increases with the increase of the potential in a cathodic direction and shows a maximum in the potential range from −100 to −300 mV. A value of 0.00 V vs Ag/AgCl is proposed as the working potential of the investigated phenol vapor biosensor because the background current of the gas biosensor is zero this potential. It should be noted that the polarization curve in the presence of phenol vapors (above 10^{-4} M phenol solution) of a gas biosensor with the same design but without enzyme in the "enzymatic layer" practically coincides with the background polarization curve shown in Figure 3.

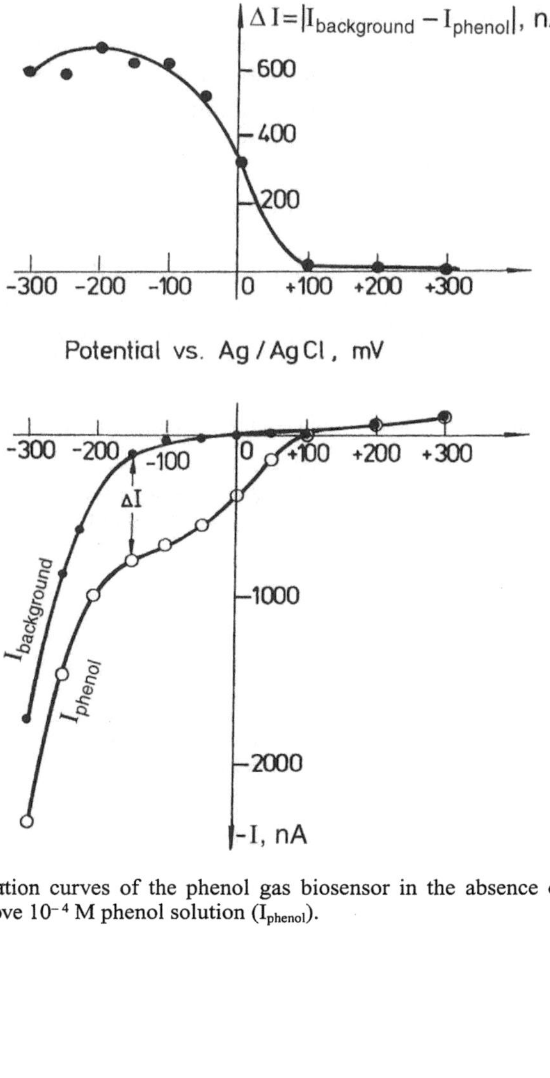

Figure 3. Polarization curves of the phenol gas biosensor in the absence of phenol vapors ($I_{background}$) and above 10^{-4} M phenol solution (I_{phenol}).

Unfortunately, in the literature there are no data for the concentration of the saturated phenol vapors which are in equilibrium with aqueous phenol solutions at very low concentrations and at room temperatures. Therefore in the abscisses of the plots presented in this paper the phenol concentrations of the aqueous solutions above which the measurement is performed are given.

However, some data can be found (Chalov, 1946) for the concentration of saturated phenol vapors in equilibrium with phenol solutions with a concentration down to 0.00079 wt% at the boiling temperature of the solution. It has been shown that in the concentration range 0.179–0.00079 wt% the ratio of the saturated phenol vapor concentration and the phenol concentration in solution (both expressed in wt%) is ca 2.32. In Table 1 some of these data are presented together with data extrapolated for lower phenol concentrations assuming that the same ratio is valid.

Figure 4 presents the calibration curve of the investigated gas biosensor obtained at a potential of 0.00 V vs Ag/AgCl by consecutive exposition of the gas biosensor above different aqueous phenol solutions. It can be seen that an amperometic signal of the biosensor is observed at a very low concentration of phenol vapor (in the ppb range – above 5×10^{-7} M phenol solution). A linear calibration curve is obtained for a wide range of phenol concentrations. The experimental points were obtained after the biosensor had been stored for 43 days at 4°C.

The gas biosensor developed on the tyrosinase/gas-diffusion electrode is sensitive also to a range of other phenolic compounds. The transient response of the biosensor to *p*-cresol vapor (when the biosensor is placed above *p*-cresol solutions) is presented in Figure 5. It can be seen that the biosensor provides a significant amperometric signal. A linear calibration

Table 1. Equilibrium gas phase concentrations above an aqueous phenol solution *at the boiling temperature:*

Concentration of phenol in the aqueous solution (M)	Concentration of phenol in the gas phase (ppm)
$3,4\times10^{-3}$	165 *)
$1,0\times10^{-3}$	42 *)
$7,0\times10^{-4}$	30 *)
$4,6\times10^{-4}$	21 *)
$3,8\times10^{-4}$	16 *)
$1,0\times10^{-4}$	4.2 **)
$8,4\times10^{-5}$	3.0 *)
$1,0\times10^{-5}$	0.42 **)
$5,3\times10^{-6}$	0.22 **)
$1,0\times10^{-6}$	0.042 **)
$5,3\times10^{-7}$	0.022 **)
$1,0\times10^{-7}$	0.0042 **)
$5,3\times10^{-8}$	0.0022 **)

*) according to Chalov (1946); **) calculated.

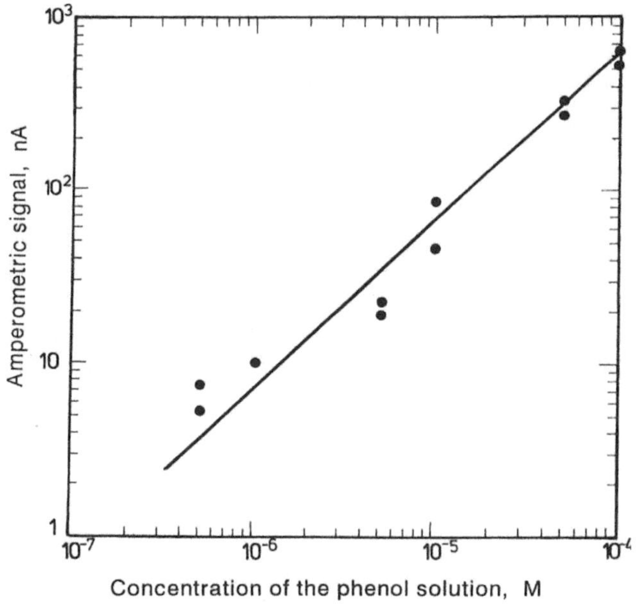

Figure 4. Calibration curve of the phenol gas biosensor obtained from the vapors above phenol solutions with different concentrations ($5 \times 10^{-7} - 1 \times 10^{-4}$ M).

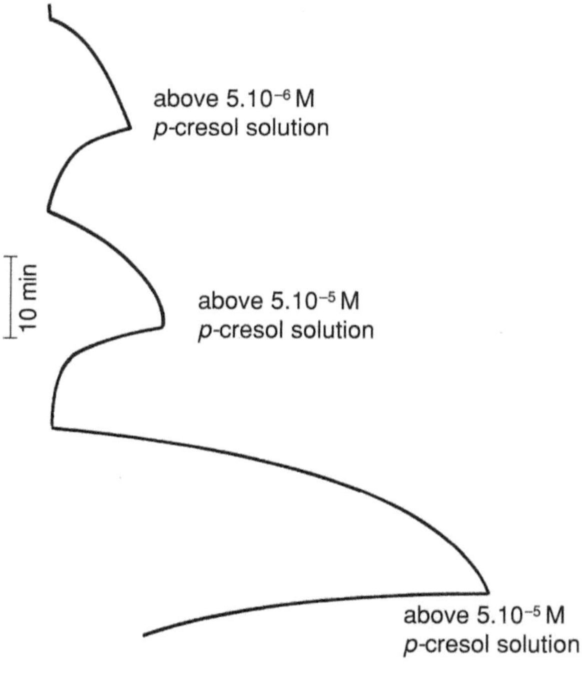

Figure 5. Transient response of a gas biosensor to vapors of *p*-cresol.

curve is obtained (Fig. 6) if the gas biosensor is placed above aqueous *p*-cresol solutions with concentrations in the range of 10^{-3} to 10^{-6} M/l.

The investigated gas biosensor is also sensitive to vapors of *4*-chlorophenol. Figure 7 displays the transient respone of the sensor when exposed above different *4*-chlorophenol solutions. The calibration curve is linear (Fig. 8) in the range of 10^{-2} to 10^{-5} M *4*-chlorophenol.

The porous membrane which separates the electrolyte in the electrochemical cell from the surrounding atmosphere simultaneously ensures an efficient exchange of molecules between the gas phase and the electrolyte through its pores. Using this principle, enzyme/gas-diffusion electrodes using suitable enzymes can probably be employed in biosensors for the detection of various gaseous substrates in the near future.

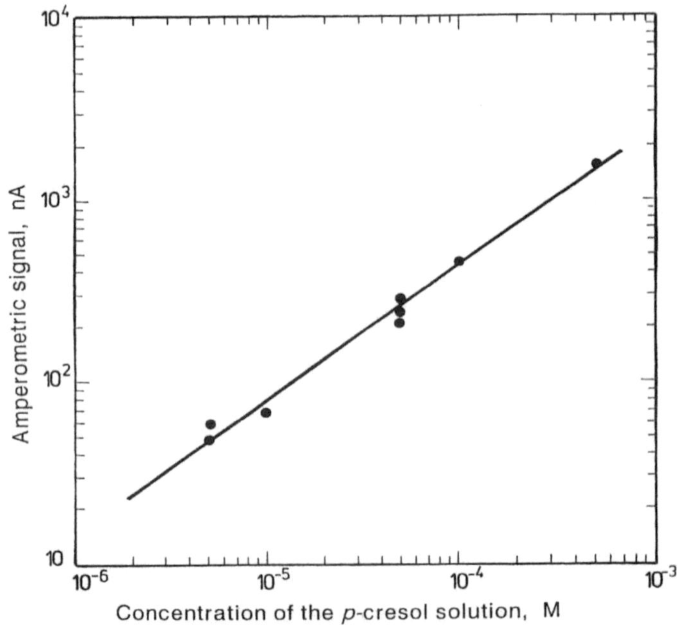

Figure 6. Calibration curve of a gas biosensor obtained from the vapors above *p*-cresol solutions with different concentrations (5×10^{-5} – 5×10^{-3} M).

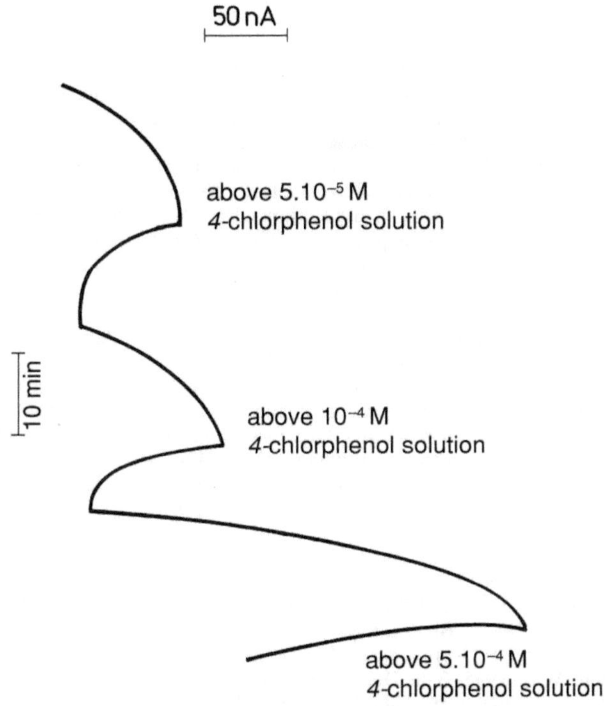

Figure 7. Transient response of a gas biosensor to vapors of *4*-chlorphenol.

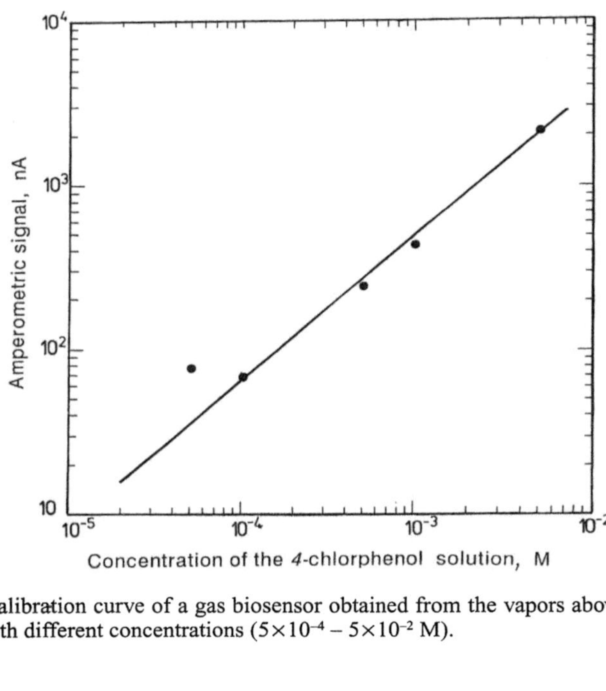

Figure 8. Calibration curve of a gas biosensor obtained from the vapors above *4*-chlorphenol solutions with different concentrations (5×10^{-4} – 5×10^{-2} M).

References

Chalov, N.V. (1946) Equilibrium between the liquid and the vapour phases in the system water – phenol for the low phenol concentrations. *Journal of Applied Chemistry* 19:1371–1372.

Dennison, M.J., Hall, J.M. and Turner, A.P.F. (1995) Gas-Phase Microbiosensor for Monitoring Phenol Vapor at ppb Levels. *Anal. Chem.* 67:3922–3937.

Dong, S. and Dasgupta, P.K. (1986) Solubility of gaseous formaldehyde in liquid water and generation of trace standard gaseous formaldehyde. *Environ. Sci. Technol.* 20:637–640.

Hämmerle, M., Hall, E.A.H., Cade, N. and Hodgins, D. (1996) Electrochemical enzyme sensor for formaldehyde operating in the gas phase. *Biosens. Bioelectron.* 11:239–246.

Iliev, I., Kaisheva, A. and Budevski, E. (1977) US Patent No 4 031 033.

Iliev, I., Kaisheva, A., Scheller, F.W. and Pfeiffer, D. (1995) Amperometric Gas-Diffusion/ Enzyme Electrode. *Electroanalysis* 7:542–546.

Kaisheva, A., Iliev, I., Kazareva, R. and Christov, S. (1995a) Amperometric enzyme/gas-diffusion electrodes. *Sensor. Actuator. B* 26–27:425–428.

Kaisheva, A., Iliev, I., Kazareva, R., Christov, S., Scheller, F.W. and Wollenberger, U. (1995b) Enzyme electrodes for determination of phenol concentration. *In: The 8th International Conference on Solid-State Sensors and Actuators, and Eurosensors IX.* Stockholm, Sweden, June 25–29, 1995, *Digest of technical papers,* Vol. 1:490–493.

Malinski, T., Taha, Z., Grunfeld, S. Burewicz, A., Tomboulian, P. and Kiechle, F. (1993) Measurements of nitric oxide in biological materials using a porphyrinic microsensor. *Anal. Chim. Acta* 279:135–140.

Mitsubayashi, K., Yokoyama, K., Takeuchi, T. and Karube, I. (1994) Gas-Phase Biosensor for Ethanol. *Anal. Chem.* 66:3297–3302.

Moseley, P.T., Norris, J.O.W. and Williams, D.E. (eds) (1991) *Techniques and mechanisms in gas sensing*. Adam Hilger, Bristol, UK.

O'Sullivan, C.K., Lafis, S., Karayannis, M.I., Dennison, M., Hall, J.M., Turner, A.P.F., Hobbs, B. and Aston, W.J. (1995) The Development Of Microbiosensors To Monitor Hazardous Gases In The Environment. *In: Fourth European Workshop on Biosensors for Environmental Monitoring*, 15–16 February, 1995, Barcelona, Spain, Proceedings, pp 44–46.

Scheller, F. and Schubert, F. (1989) *Biosensoren*. Akademie-Verlag, Berlin.

Sigrist, M.W. (ed.) (1994) Air monitoring by spectroscopic techniques. *In: Chemical Analysis,* Vol. 127. John Wiley & Sons, New York.

Vairavamurthi, A., Roberts, J.M. and Newnan, L. (1992) Methods for determination of low molecular weight carbonyl compounds in the atmosphere: a review: *Atmosph. Environ.* 26A: 1965–1993.

Frontiers in Biosensorics II
Practical Applications
ed. by. F. W. Scheller, F. Schubert and J. Fedrowitz

Microbial BOD sensors: Problems of practical use and comparison of sensorBOD and BOD_5

K. Riedel

Dr. Bruno Lange GmbH Berlin, Industriemeßtechnik, 40549 Düsseldorf, Germany

Summary. For some years microbial sensor system for the determination of BOD have been use in waste water control. The main advantage of the microbial BOD sensor is the short measuring time. However, the sensorBOD-values are not identical with BOD_5 in all cases. These differences are caused by the different measuring principles and the variable composition of waste waters. The microbial sensor gives a response on the easily assimilable compounds in waste water. Possibilities to improve the coincidence of sensorBOD and BOD_5 are: (i) selection of microorganisms with suitable substrate spectrum and combination of microbial species, (ii) alteration of sensor activity by pre-incubation with the desired substrate or waste water, (iii) enzymatic or acid hydrolysis of waste water, and (iv) use of suitable calibration substrate.

Introduction

The difficulties involved in analyzing the numerous substances that are present in a large range of concentrations in waste water make sum parameters an indispensable part of waste water monitoring systems. A parameter of particular importance is *Biochemical oxygen demand* (BOD), because it allows conclusions to be drawn about the biodegradable substrate. At present, the main disadvantage of this parameter is the time of 5 days necessary to obtain the result. Therefore the conventional BOD is unsuitable for process control. A more rapid estimation of BOD is possible by using microbial sensors containing whole cells immobilized on an oxygen electrode.

The first report of such microbial BOD sensor was published by Karube et al. in 1976. Since 1983 a BOD-sensor system based on this microbial BOD-sensor has been produced by Nisshin Electric Co. Ltd. (Karube, 1986). Meanwhile, commercially available BOD-sensor systems are being produced by AUCOTEAM GmbH Berlin (Merten and Neumann, 1992), PGW GmbH Dresden (Szweda and Renneberg, 1994), and Dr. Bruno Lange GmbH Berlin (Riedel et al., 1993).

Construction and function

The design of the microbial BOD sensor shown in Figure 1 is in principle identical to an enzyme sensor. The main parts of such a biosensor are the immobilized microorganisms as a biological recognition system and an oxygen electrode as physical transducer. The parts are separated by a gas

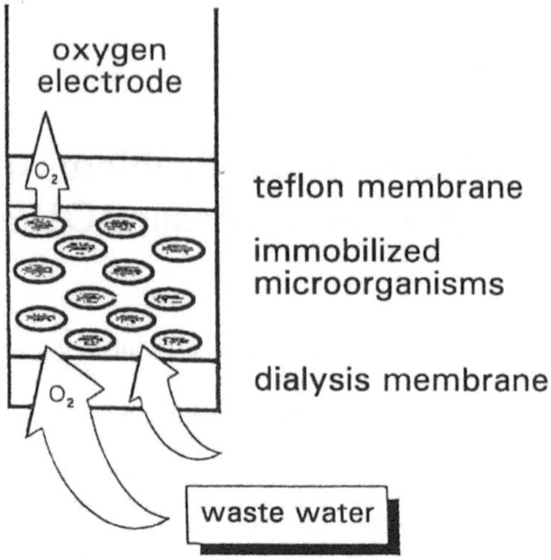

Figure 1. Schema of microbial BOD sensor.

permeable membrane. The cells are immobilized using an outer semipermeable membrane covering the sensor.

Microorganisms

The prerequisite for the use of microorganisms for BOD sensors is a wide substrate spectrum or, in other words, the "multireceptor" behaviour enabling the recognition of a group of substances. Karube et al. (1977) and Strand and Carlson (1984) have been using activated sludges obtained from waste water treatment plants. However, such biosensors with an undefined variety of microbial species did not give reproducible results (Hikuma et al., 1979). Sensors using pure cultures of microorganisms seem to be more suitable. BOD sensors have been developed using various microorganisms: *Trichosporon cutaneum* (Hikuma et al., 1979; Harita et al., 1985; Riedel et al., 1988; 1990), *Hansenula anomala* (Kulys, 1980; Li and Chu, 1991). *Issatchenkia orientalis* (Riedel et al., 1993; 1994; 1997), *Torulopsis candida* (Rajasekar et al., 1992), *Pseudomonas putida* (Ohki et al., 1990), *Rhodococcus erythropolis* (Riedel et al., 1993; 1994; 1997), *Bacillus licheniformis* (Tan et al., 1992); *Bacillus polymyxa* (Su et al., 1986) and *Bacillus subtilis* (Riedel et al., 1988; Tan et al., 1992).

Transducers

The most commonly used transducer of BOD sensors is the amperometric oxygen electrode. The application of an optical oxygen transducer instead of the usual amperometric oxygen electrode was described by Preininger et al. (1994). Another interesting possibility is the use of the luminous bacteria *Photobacterium phosphoreum* (Hyan et al., 1993). The background of this device is the relation of intensity of luminescence to the cellular assimilation of organic compounds in waste water.

The use of a biofuel cell type electrode for BOD determination has been described by Karube et al. (1977). The current generated by the biofuel cell resulted from the oxidation of hydrogen and formate produced from organic compounds by *Clostridia* under anaerobic conditions.

Function of microbial BOD sensor

BOD determination with a microbial sensor is based on direct measurement of the oxygen consumption of the microorganisms at the oxygen electrode. This consumption depends on the organic substrates in the sample.

If a waste water sample is added to the measuring system the biodegradable substrates are used by the microorganisms of the sensor for respiration, causing an immediate decrease in the oxygen concentration. This effect is measured with the oxygen electrode.

In principle, there are two possibilities for measurement: (i) end-point measurement (where the difference of current, I, reflects the respiration rate of the substrates), and (ii) the kinetic measurement (first derivative of the current-time curve). The first method has been most frequently used in microbial sensors, for example by the BOD-sensor system of Nisshin electric (Karube, 1986). Here a relatively high concentration of biomass and thick membranes are used.

The short measuring period, which is used in the instruments by PGW GmbH Dresden (Szweda and Renneberg, 1994), AUCOTEAM GmbH Berlin (Merten and Neumann, 1992) and Dr. Bruno Lange GmbH Berlin (Riedel et al., 1993), has a positive effect of the measuring frequency and stability of the sensor. Owing to the short exposure time of no more than 1 minute the accumulation of substrates in the microbial cells of the sensor remains at a low level. This is important because the sensor is not ready to carry out another measurement until the microorganisms are again hungry. Moreover the stability of the sensor is considerably increased. Any toxic substances that may be present in the waste water act on the sensor for a short period only.

Application of BOD sensor

Comparison of sensorBOD and BOD_5

The sensorBOD systems were used to determine the BOD of waste water flowing into and out of municipal and industrial sewage treatment plants of

Table 1. BOD-values estimated by microbial sensors and determined by the 5-day method for various waste water samples.

Waste water	Microbial sensor	BOD_5	SensorBOD	BOD_5-sensorBOD-ratio	References
Municipal waste water	*B. subtilis* + *B. licheniformis*	170	154	1.10	Tan et al., 1992
	Issatchenkia + *Rhodococcus*	285	334	0.85	Riedel et al., 1997
	Issatchenkia + *Rhodococcus*	366	366	1.00	
	Issatchenkia + *Rhodococcus*	131	90	1.46	
	Issatchenkia + *Rhodococcus*	180	162	1.11	
	Issatchenkia + *Rhodococcus*	112	63	1.78	
	Issatchenkia + *Rhodococcus*	53	23	2.30	
	Issatchenkia + *Rhodococcus*	108	50	2.16	
	Issatchenkia + *Rhodococcus*	153	57	2.68	
	Issatchenkia + *Rhodococcus*	114	68	1.68	
	Issatchenkia + *Rhodococcus*	166	100	1.66	
	Issatchenkia + *Rhodococcus*	169	130	1.30	
Food factory	*Trichosporon cutaneum*	152	155	0.98	Riedel et al., 1990
	Trichosporon cutaneum	8000	8764	0.91	
	B. subtilis + *B. licheniformis*	151	147	1.03	Tan et al., 1992
Starch factory	*Trichosporon cutaneum*	4000	4250	0.94	Riedel et al., 1990
Fermentation factory	*B. subtilis* + *B. licheniformis*	15040	15640	0.96	Tan et al., 1992
Industrial waste water	*Issatchenkia* + *Rhodococcus*	253	389	0.65	Riedel et al., 1997
	Issatchenkia + *Rhodococcus*	400	677	0.59	

various sizes, and the values obtained were compared with the conventional BOD_5. As demonstrated in Table 1, for domestic waste water the sensor often showed low values compared with BOD_5. Comparable estimates for BOD were obtained for untreated waste water from fermentation and food plants. The sensorBOD values are, however, not identical with BOD_5 in all cases. The cause of this behaviour is the variable composition of waste water. The microbial sensor gives a response on the easily assimilable compounds in waste water. Waste water with a high content of easily assimilable compounds, such as that from fermentation and food plants, gives comparable values for BOD by sensor and conventional method. The sensor showed low values compared with BOD_5 for waste water, e.g. domestic samples, containing relatively high concentrations of polymers, such as proteins, starch and lipids, and low concentrations of easily assimilable compounds.

Therefore, the BOD sensor is well applicable to specific waste waters generated by the food and fermentation industry with a high content of easily assimilable compounds; however, its applicability to domestic waste water is limited (Tanaka et al., 1994).

Differences between sensorBOD and BOD_5

The oxygen demand of a sample is measured in a conventional test during 5 days reflecting the various metabolic reactions of a mixed population. In other words, the BOD_5 measures the sum of various biochemical processes in a biosludge over a period of five days, including the adaptation to substrates by the induction of degradation enzymes and the enzymatic hydrolysis of polymers, e.g. starch, proteins, lipids. By contrast, the determination of BOD with a microbial sensor is a fast test of biological activity with a selected microbial species, providing a sort of snapshot. Therefore, polymers can not be determined with a BOD sensor, because the hydrolysis of these polymers in the short measuring time of the sensor is impossible. But it is possible to achieve the metabolic reactions and alterations of microorganisms, which take place by BOD_5 over a period of 5 days, by considering the following pretreatments of microbial sensor and waste water: (i) induction of metabolic capacities by incubation of the biosensor with the desired substances or waste water and (ii) hydrolysis of polymers by enzymatic or acid treatment of waste water. Further possibilities to improve the coincidence of BOD sensor and BOD_5 value for domestic effluent can be achieved by the selection of appropriate microorganisms and the use of "artificial" waste water as a modified calibration standard.

Moreover, these problems are not encountered when using a BOD sensor for process control, because in this case easily assimilable substrates are used.

Improvement of coincidence of sensorBOD and BOD_5

Selection of microorganisms with suitable substrate spectrum and combination of microbial species

A fundamental distinction between the microbial BOD sensor and the 5-day method is that the biosensor uses a single species of microorganisms, whereas the conventional method uses many microorganism species obtained from activated sludge of a waste water plant. One microbial species has a specific substrate spectrum and it can determine only a portion of the organic compounds of waste water. In this case the BOD value is, in comparison to the BOD_5 method, too small. A possibility to improve the coincidence with BOD_5 is the selection of appropriate microorganisms, as described by Tanaka and coworkers (1994). Another way is the combination of microorganisms with a different substrate spectrum which leads to a biosensor with improved quality. Mixed cultures of the

Table 2. Sensitivity of BOD sensor containing *Rhodococcus erythropolis* or *Issatchenkia orientalis* as well as their combination in comparison to BOD (Calibration with glycerol) (Riedel et al., 1997).

Substrate	BOD [mg/mg] (Bund and Straub, 1973)	Sensor BOD [mg/mg]		
		Rhodococcus	*Issatchenkia*	*Rhodococcus* + *Issatchenkia*
Glycerol	0.8	0.8	0.80	0.80
Fructose	0.6	2.77	0.19	0.90
Glucose	0.6	0.51	0.56	0.31
Sucrose	0.7	0.21	0.07	0.06
Maltose	0.7	0.06	0.20	0.07
Lactose	0.55	0.03	0.01	0.02
Acetic acid	0.35	4.46	3.44	3.65
Lactic acid	0.63	1.56	0.70	0.22
Ethanol	1.5	22.04	1.08	6.28
Glutamine acid	0.64	2.24	1.33	2.04
Glycine	0.52	0.13	0.50	0.29
Alanine	0.94	0.13	0.66	0.25
Tryptophan		0.02	0.10	0.07

closely related strains were used by Tan et al. (1992) (*B. subtilis* and *B. licheniformis*) as well as Galindo et al. (1992) (*Enterobacter* spec. and *Citrobacter* spec.). The sensor with the bacteria *Rhodococcus erythropolis* and *Issatchenkia orientalis* associates the substrate sensitivity of both strains, as demonstrated in Table 2 (Riedel et al., in press). This combination of two strains with different substrate specificities allows the construction of a BOD sensor, providing an adequate value to BOD_5.

Alteration of sensor activity by pre-incubation with the desired substrate or waste water

A possible increase of BOD sensor sensitivity is obtained by the pretreatment of the sensor with the desired substrates. As shown in Table 3, comparative results were obtained for BOD values estimated by pre-incubated microbial sensor and those determined by the conventional BOD method.

Enzymatic or acid hydrolysis of waste water

Waste water with a high content of polymers, such as proteins, starch or cellulose, gives biosensor BOD-values, which are smaller than the BOD values resulting from the 5-day method. The cause is that the microbial sensor is not able to determine these compounds. It is possible to split these compounds by an enzymatic pretreatment of waste water analogous to what happens during the 5-day conventional BOD test. The effect of enzymatic treatment of waste water on the sensorBOD is demonstrated in the example of a paper factory. Such waste water gives biosensor BOD-values, which are smaller than the BOD resulting from the 5-day method. The cause is that the microbial sensor is not able to determine cellulose. A drastic

Table 3. Increase of sensor activity by adapting an appropriate substrate or waste water incubation.

Substrate Waste water	BOD sensor	BOD_5 [mg/l]	Sensor BOD [mg/l]		References
			untreated	induced	
Fructose*	*Trichosporon*	0.71	0.44	0.60	Riedel et al.,
Maltose*	*cutaneum*	0.49–0.76	0.02	0.15	1990
Alanine*		0.55	0.19	0.41	
OECD model waste water		18000	10640	13680	
Chemical factory		726	470	853	
Food factory		8000	4076	8764	
Paper factory	*Issatchenkia*	n.d.	91	146	Riedel et al.,
Municipal	+*Rhodococcus*	n.d.	15	46	1997
waste water		n.d.	14	34	
Industrial waste water					

* mg BOD/mg substance.

Table 4. BOD-values determined after pretreatment of waste water from a paper factory with cellulases and β-glucosidases.

Waste water	Sensor BOD [mg/l]		Increase of BOD [%]
	untreated	enzymatic treatment	
Sample 1	10	29	290
Sample 2	12	22	183
Sample 3	44	75	170
Sample 4	26	39	150
Sample 5	5	21	420

increase of sensor BOD of such waste water is achieved by treatment with enzymes (Table 4) or acid. The polymers in waste water are split into their monomers such as amino acids, monosaccharides, glycerol and fatty acids, by acid hydrolysis at 1 h and 100 °C (Kasel et al., 1996). The hydrolysis of waste water caused in most cases an increase of BOD as shown in Table 5. The concurrence of sensorBOD and BOD_5 is improving (Riedel et al., 1997).

Calibration and calculation of BOD_5

The sensorBOD is a biological activity test, i.e. the organic substances in waste water are determined from the respirational changes of the microorganisms of the sensor. Like all biological activity tests the microbial BOD sensors need to be calibrated before a comparison can be made with the conventional BOD.

Table 5. Sensor BOD-values of untreated and hydrolyzed domestic waste water in comparison with 5-day BOD and chemical oxygen demand (COD) (Riedel et al., 1997).

Waste water Sample-No.	BOD_5 [mg/l]	COD [mg/l]	Sensor BOD [mg/l]		Sensor BOD_5 ratio	
			untreated	hydrolyzed	untreated	hydrolyzed
MW6	131	370	90	85	0.69	0.65
MW8	123	331	91	103	0.74	0.84
MW11	180	469	162	160	0.90	0.89
MW12	3.3	60	6	7	1.82	2.12
MW13	112	275	63	103	0.56	0.92
MW14	1	38	1	1	1.00	1.00
MW15	53	201	23	31	0.43	0.58
MW17	108	264	50	63	0.46	0.58
MW18	153	572	57	110	0.37	0.72
MW21	114	256	68	77	0.60	0.68
MW22	166	761	100	113	0.60	0.68
MW23	169	605	130	175	0.77	1.04
MW24	3.5	18	2	3	0.57	0.86

Table 6. Comparison between BOD estimated by the microbial sensor (using *Issatchenkia* and *Rhodococcus*) and determined by 5-day method (Riedel et al., 1997)

Substrate	Sensor BOD-BOD_5 ratio
Glucose	0.31
Fructose	0.90
Sucrose	0.06
Lactose	0.02
Maltose	0.07
Glycerol	1.00
Acetate	3.65
Lactate	0.22
Ethanol	6.28
Alanine	0.28
Glycine	0.29
Glutamic acid	2.04

For calibration of a BOD sensor the so-called GGA-standard (glucose and glutamic acid) according to BOD_5-standards (JIS K 3602) (1990), glucose or glycerol with predetermined BOD_5 (Riedel et al., 1990, 1994), can be used.

The usually used GGA-standard for BOD_5-calibration is not suitable for microbial sensors, because (i) this standard is instable due to microbial contamination, and (ii) the glutamic acid reaction of microorganisms is decreased in the presence of glucose due to glucose repression.

A problem is that the response of microbial sensors to pure substances is different from that to the reference substance, as the sensorBOD/BOD_5-factor in Table 6 shows. Compounds such as lactose, maltose, sucrose give lower values than when determined by the 5-day test. Conversely, for acetic acid and ethanol, the sensorBOD-values are higher than the BOD_5-values. Therefore Tanaka et al. (1994) proposed as standard an artificial, treated waste water containing beef extract, peptone, nitrohumic acid, tannic acid, lignin sulfuric acid, sodium lauryl, sulfate gum arabic and minerals. This standard is somewhat unsuitable, because it is not stable.

Under defined conditions it is possible to calculate the BOD_5 from the sensorBOD-values with the help of specific conversion factors. This requires that the qualitative composition be relatively constant and only the quantitative composition is altered. Moreover, the content of easily assimilable compounds must be high in comparison to polymers. These factors fave a specific character and any given factor is applicable only to one particular stage of an individual sewage treatment plant.

Conclusion

The short measuring time and high precision of the sensorBOD system open up novel opportunities in the field of waste water monitoring, and give dischargers and sewage treatment plant operators the means to comply with steadily increasing demands to carry out their own checks. However, the sensorBOD values are not identical with BOD_5 in all cases. The reasons for differences are the different measuring principles and the variable compositon of waste waters. The microbial sensor gives a response to the easily assimilable compounds in waste water. Therefore the sensorBOD is a new parameter in waste water monitoring. It is necessary to recognize this parameter as standard, as in Japan (JIS K 3602, 1990). At present the regulations of the German Länder on in-house monitoring of waste water state that a simplified or alternative method for determining the BOD can in principle be used if the measured value obtained with the method reliably fulfils the specified aim of the analysis.

Official acceptance of the sensorBOD requires proof of a correlation between the sensorBOD and the legally binding BOD_5 determined by means of the conventional method. A calculation of BOD_5 from sensorBOD with a factor is in many cases possible. This requires that the qualitative composition remain relatively constant and only the quantitative composition be altered.

Moreover, the content of easily assimilable compounds must be high in comparison to polymers. Finally it is important that the coincidence of sensorBOD and BOD_5 be improved by the selection of microorganisms with broad substrate spectrum.

References

Bond, R.G. and Straub, C.P. (1973) Handbook of environmental control. Vol. 3. Cleveland, Ohio, pp 671–686.

Galindo, E., Garcia, J.L., Torres, L.G. and Quintero, R. (1992) Characterization of microbial membranes used for the estimation of biochemical oxygen demand with a biosensor. *Biotechnol. Techniques* 6:399–404.

Harita, K., Otani, Y., Hikuma, M. and Yasuda, T. (1985) BOD quick estimating system utilizing a microbial electrode. *In:* R.A.C. Drake (ed.): *Instrum. Control. Water Wastewater Treat. Transp. Syst.* Proc. IAWPRC Workshop, pp 529–532.

Hikuma, M., Suzuki, H., Yasuda, T., Karube, I. and Suzuki, S. (1979) Amperometric estimation of BOD by using living immobilized yeasts. *Eur. J. Appl. Microbiol. Biotechn.* 8:289–97.

Hyan, C.-K., Tamiya, E., Takeuchi, T. and Karube, I. (1993) A novel BOD sensor based on bacterial luminescence. *Biotechnol. Bioeng.* 41:1107–1111.

JIS K 3602 (1990) Japanese Industrial Standard: Apparatus for the estimation of biochemical oxygen demand (BOD_5) with microbial sensor.

Karube, I. (1986) Trends in bioelectronics research. *Science Technol. Japan.* July/Sept. 32–40.

Karube, I., Matsunaga, T., Mitsuda, S. and Suzuki, S. (1977) Microbial electrode BOD sensor. *Biotechn. Bioeng.* 19:535–547.

Kasel, A., Grabert, E. and Frischwasser, H. (1996) Studies on the measurement of the Biochemical Oxygen Demand (BOD): sensor-BOD in comparison with BOD_5 according to DEV H 51. BIOspektrum: PEO32.

Kulys, J. and Kadziauskiene, K. (1980) Yeast BOD sensor. *Biotech. Bioeng.* 22:221–226.

Li, Y.R. and Chu, J. (1991) Study of BOD microbial sensors for waste water treatment control. *Appl. Biochem. Biotechnol.* 28:855–863.

Merten, H. and Neumann, B. (1992) BSB-Kurzzeitmessung mit Biosensor. *BioTec* 6:43–46.

Ohki, A., Shinohara, K. and Maeda, S. (1990) Biological Oxygen demand sensor using an arsenic resistant bacterium. *Anal. Sci.* 6:905–906.

Preininger, C., Klimant, I. and Wolfbeis, O.S. (1994) Optical fiber sensor for biological oxygen demand. *Anal. Chem.* 66:1841–1846.

Rajasekar, S., Madhav, V.M., Rajasekar, R., Jeyakumar, D. and Rao, G.P. (1992) Biosensor for the estimation of biological oxygen demand based *Torulopsis candida*. *Bulletin of Electrochem.* 8:196–198.

Riedel, K. and Uthemann, R. (1994) SensorBSB – neuer mit Biosensoren gewonnener Summenparameter in der Abwasseranalytik. *Wasserwirtsch.-Wassertechn.* 2:35–38.

Riedel, K., Renneberg, R., Kühn, M. and Scheller, F. (1988) A fast estimation of BOD with microbial sensors. *Appl. Microbiol. Biotechn.* 28:316–318.

Riedel, K., Lange, K.-P., Stein, H.J., Kühn, M., Ott, P. and Scheller, F. (1990) A microbial sensor for BOD. *Water Res.* 24:883–887.

Riedel, K., Kloos, R. and Uthemann, R. (1993) Minutenschnelle Bestimmung des BSB. *LWB Wasser, Boden und Luft* 11–12:35–38.

Riedel, K., Uthemann, R.. Yang, X. and Renneberg, R. (1997) Determination of BOD in waste water with a commercial combination-sensor containing *Rhodococcus erythropolis* and *Issatchenkia orientalis*. *Biosens. Bioelectron.* 12: in press.

Strand, S.E. and Carlson, D.A. (1984) Rapid BOD measurement for municipal wastewater samples using a biofilm electrode. *J. Water Pollut. Control Fed.* 56:464–467.

Su, Y.C., Huang, J.H. and Liu, M.L. (1986) A new biosensor for rapid BOD estimation by using immobilized growing cell beads. *Proc. Natl. Sci. Counc. B. ROC* 10:105–112.

Szweda, R. and Renneberg, R. (1994) Rapid BOD measurement with the Medingen BOD-module. *Biosens. Bioelectron.* 9:IX–X.

Tan, T.C., Li, F., Neoh, K.G. and Lee, Y.K. (1992) Microbial membrane-modified dissolved oxygen probe for rapid biochemical oxygen demand measurement. *Sensor. Actuator. B* 8:167–172.

Tanaka, H., Nakamura, E., Minamiyama, Y. and Toyoda, T. (1994) BOD biosensor for secundary effluent from wastewater treatment plants. *Wat. Sci. Tech.* 30:215–227.

Frontiers in Biosensorics II
Practical Applications
ed. by. F.W. Scheller, F. Schubert and J. Fedrowitz

New biosensors for environmental analysis

P.-D. Hansen and A. v. Usedom

Berlin University of Technology, FB 7, Institute for Ecological Research and Technology, Department of Ecotoxicology, D-10589 Berlin, Germany

Summary. Environmental analysis requires fast and reliable measurement results. Biosensors, which facilitate integral monitoring as well as single substance analysis, achieve high sensitivities in a minimum of measuring time. Four new on-line biosensors, which cover a wide range of environmentally relevant substances, are introduced: A water-quality monitoring bacteria electrode, whose gradual development is described as an example, a heavy-metal screening urease inhibition sensor, a genotoxic potential as well as a immunotoxic potential indicating sensor. Future prospects are given.

Significance of biosensors for environmental analysis

Breakdowns in chemical and pharmaceutical industry, which sometimes caused serious loading rates of river and drinking water, made clear how necessary early warning systems have become in environmental analysis. Until now, pollutant loadings were only indicated by changes in the ecosystem, e.g. reduction of occurring species, mutations, death of fish. New systems, which are able to detect on-line a variety of pollutants in a minimum of time, are in demand. A promising way is the use of biosensors.

Which kind of biological component would be useful as an alarm transmitter? Enzymes and antibodies are well known for their selectivity of interaction with the analyte and therefore mainly used in single substance analysis. However, unicellular organism as well as organelles are suitable for integral water-quality monitoring because of their wide-band sensitivity for toxic agents. The type of application determines the choice of sensor material.

The goals of our investigations were the development of efficient environmental biosensors – each of them determining different families – and their following combination in a cluster technology.

Thus, on the one hand we constructed an amperometric sensor for water-quality monitoring, the development of which is described below. Bacteria such as *Synechococcus leopoliensis* and *Escherichia coli* (Hansen et al., 1989; Hansen, 1992b; Hansen and Stein 1994; v. Usedom, 1995), as well as algae, marine and non-marine, are used to indicate mainly *pesticides.* On the other hand, a way for *heavy metal* detection was realized: The inhibition of urease activity (Wittekindt et al., 1996) is the principle of this on-line biosensor. Finally, under the regulations of German water legislation, the *genotoxic potential* (Hansen, 1995; Rao et al., 1995) as well as the *immunotoxic potential* (Hansen et al., 1991; Hansen, 1992a) can be determined by new sensor systems based on umu-assay and phagocytosis activity, respectively.

Measuring results, obtained from these biosensor multiarrays, form the basis for a reliable ecotoxicologic assessment of the observed system. They are applied to the protection of drinking water and surface water, but also to the monitoring of effluent streams of industrial plants, sewage plants, etc.

Amperometric biosensor for aquatic pollutant detection

Methodology

The measuring principle is to record the biochemical electron transport system (ETS). To achieve this, living organisms are fixed to a working electrode, which is exposed to the water sample. Additional redox-mediators, providing the electrochemical linkage between organisms and electrode surface, are reduced either by the photosynthetic (*Synechococcus*) or the respiratory (*E. coli*) chain (Hansen and Stein, 1994). The application of a constant potential results in the re-oxidation of the mediator at the working electrode.

Constant intervals of switching the light and mediator supply, respectively, form current maxima and minima, which allows for an evaluation independent of absolute current values.

Harmful substances, as e.g. pesticides, cause lower vitality of the organism and consequently reduced signals. This inhibition of the ETS is a measure for the toxic load of the water sample.

Several biosensors of this type have been constructed and commercialized. One of them is in operation at the water-quality monitoring station Kleve/Bimmen, Germany, at the river Rhine (Hansen and Stein, 1994). Pesticides can be detected in a sensitivity range of < 10 µg/l (e.g. atrazine).

The main goals of our latest optimization of the sensor, carried out with *Synechococcus*, were a further increase of its sensitivity, simplification of its handling, as well as a higher dwelling time and ruggedness (v. Usedom, 1995).

The obtained results were implemented into a flow-through system for on-line monitoring as well as into a single-rod electrode for immediate toxicologic measurements.

In the following some aspects of the gradual optimization of the bacteria electrode are briefly introduced.

Results of experimental investigations

Cultivation and storage of the organisms

Synechococcus was cultivated in batch flasks ($V = 0.5$ l) at a temperature of 20 °C in Blue Green 11 medium. The light supply was carried out by universal-white fluorescent lamps (1000 lx), switching dark (17h) and light (7 h) intervals.

Algae from the exponential growth stage were immobilized on filter-plates, dried and stored for many months at room temperature without any loss of vitality. Freezing these filter-plates at – 80°C even resulted in an increase of photosynthetic activity.

Type and concentration of the mediator

In the system described in the chapter on methodology, $K_3[Fe(CN)_6]$ was dispensed to the flow system. The optimal concentration of mediator was determined (5 mmol/l); an alternative mediator, ferrocene, was also examined.

As a water-insoluble mediator, ferrocene was fixed at the electrode surface by dip-coating with chloroform. In this way, multiple usuability of the water sample as well as cost reduction became possible.

The use of $K_3[Fe(CN)_6]$ leads to ten times higher current values than the use of ferrocene (see Fig. 1).

Although the application of $K_3[Fe(CN)_6]$ is preferable, the feasibility of using ferrocene as a mediator is demonstrated.

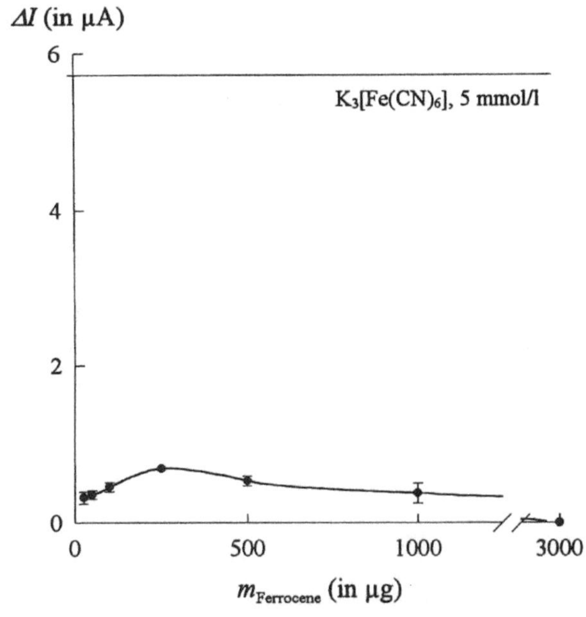

Figure 1. Influence of mediator concentration. Ferrocene, immobilized together with *Synechococcus* on the electrode surface (∅ = 5 mm). The current value measured in 5 mmol/l $K_3[Fe(CN)_6]$, carried out by the same algae, is given as a reference.

Process of immobilization
Filter-plate immobilization of *Synechococcus*, as already used in the former system, is easy, cheap and more efficient than other methods, e.g. fixing membranes or polymers. Discs of 5 mm diameter are held on the electrode surface by a nylon network. Dwelling times up to one week were achieved in this way.

Light supply of the algae
Different kinds of light supply were compared. For monochromatic light, a wavelength of 660 nm is optimal.

Cyanophycea are able to use light of wavelengths between the two absorption maxima of chlorophyll a (660 and 470 nm). Therefore a mix of red light (660 nm) and additional green light (565 nm) results, at the same total light intensity, in a higher photochemical yield and consequently in higher current values than monochromatic light of 660 nm.

The new sensor

Flow-through cell
Based on the experiences from the batch system, a flow-through cell was constructed (Fig. 2).

The calibration curve in Figure 3 was obtained by measurements with the herbicide atrazine as a model substance. The lowest observed effect concentration (LOEC) is 5 µg/l.

Single-rod electrode
A single-rod electrode for immediate toxicologic measurements was constructed and is shown in Figure 4. Its sensitivity range is the same as that of the flow-through cell.

In case of a breakdown with suspected water contamination, a reference value is measured in a water matrix similar to the real water. The difference to the value of the real water sample is a measure for the toxic load of the observed system. Thus, a reliable assessment of toxicity is possible within a few minutes.

Biosensor for heavy metal detection

The sensor for the detection of heavy metals is derived from a microtiter plate based urease inhibition assay (Wittekindt et al., 1996). The test takes advantage of the fact that many heavy metals inhibit the activity of the enzyme urease. This is measured by detecting the amount of ammonia produced with and without the presence of sample. The larger the reduction of produced ammonia, the higher the amount of heavy metals. Ammonia is

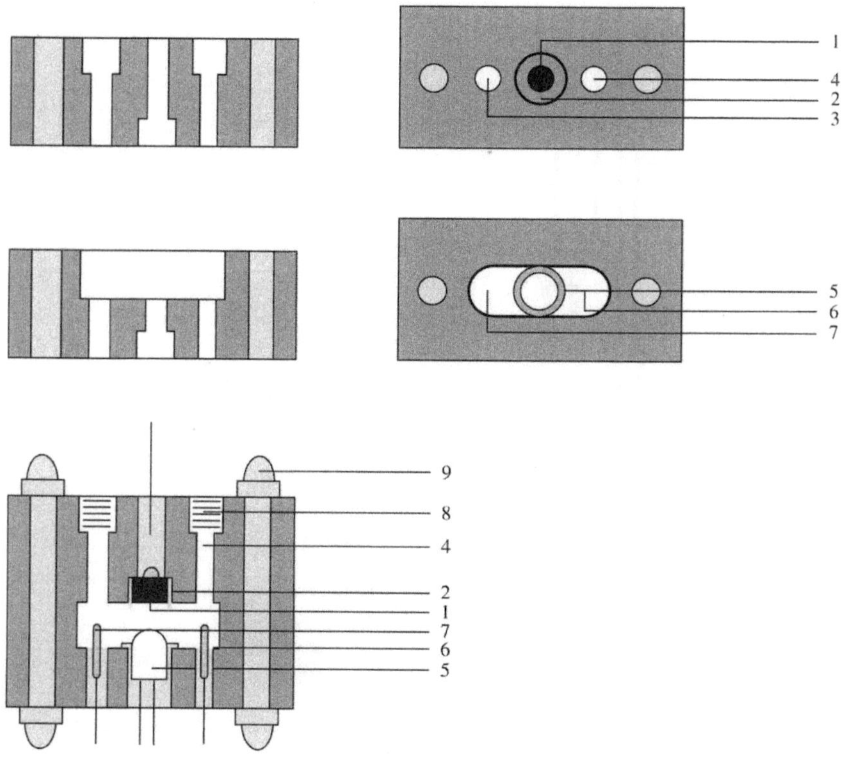

Figure 2. Flow Through cell (1:1). 1 Pt ; 2 plastic cap; 3 run in; 4 run out; 5 LED, Ga/Al/As, 660 nm, 3000 mcd, 20 mA; 6 and 7 Ag/AgCl-reference; 8 UNP-thread for FIA connectors; 9 screws of brass.

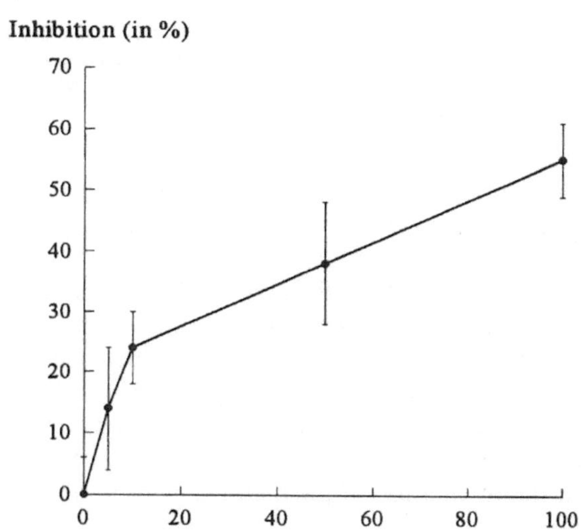

Figure 3. Inhibition of algae vitality by atrazine.

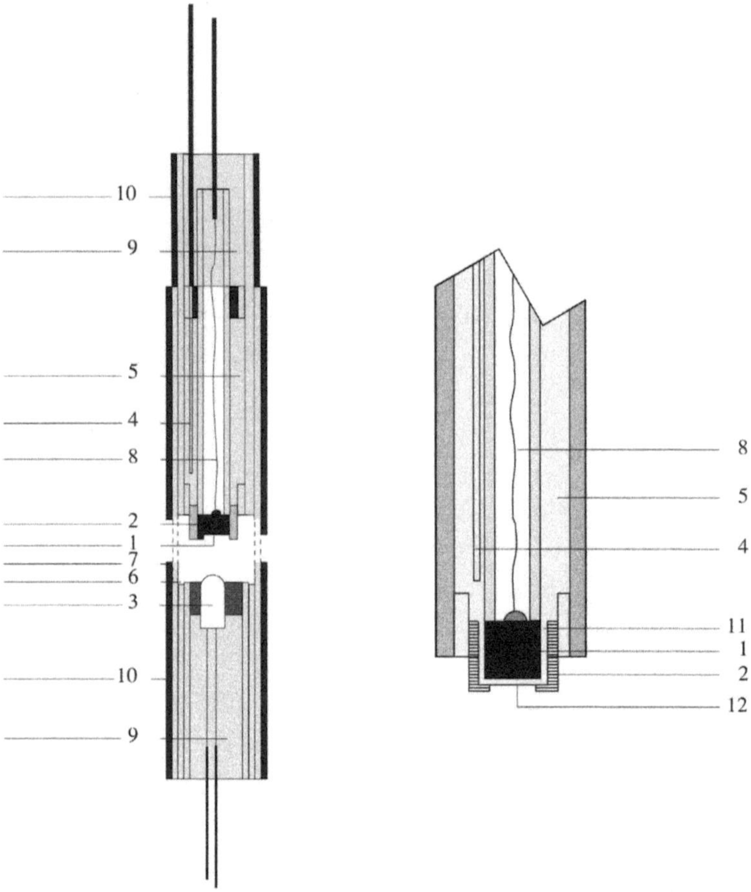

Figure 4. Single-rod electrode (1:1). 1 Pt; 2 plastic cap; 3 LED, Ga/Al/As, 660 nm, 3000 mcd, 20 mA; 4 Ag/AgCl-reference; 5 mediator solution (0.1 mol/l); 6 sample area; 7 sample run in; 8 silver wire (0.25 mm); 9 acrylic glass; 10 black coating; 11 capillar gap (0.2 mm); 12 filter plate with algae.

detected photometrically, by reaction with a chromogenic substance, or electrochemically, e.g. by an ion sensitive electrode. Another pathway is the detection of the pH-value.

The microtiter plate assay is transformed, on the one hand, into an automated on-line assay, using dissolved urease. On the other hand, an on-line biosensor with the enzyme cross-linked to a glass electrode surface (Meier et al., 1992), measuring the pH-value, is constructed.

The working systems are finally validated by using heavy metal standards. The urease detection system will be used for screening of heavy metals by the sensor. For the quantification of the metal species antibodies to Hg, Cd and Pb are available.

The flow scheme of the heavy metal sensor is presented in Figure 5.

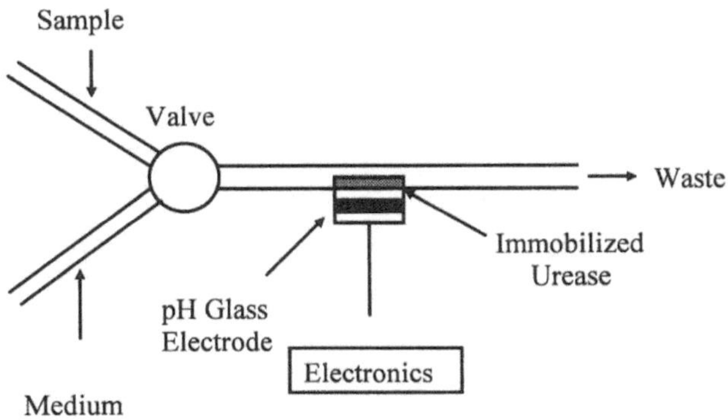

Figure 5. Flow scheme of the urease inhibition sensor.

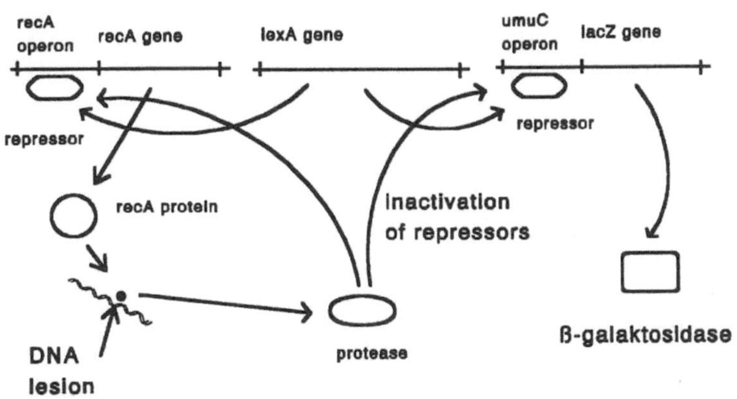

Figure 6. Principle of the umu-assay.

Genotoxicity sensor

The sensor for the detection of the genotoxic potential in waterways (Hansen, 1995) and effluent testing (ISO 1996) is based on the umu-assay after Oda et al. (1985) (see Fig. 6), which is currently performed on the microtiter plate. The principle of the umu-assay is the bacterial response to genotoxic effects. In this case the SOS-repair system is activated. In the bacteria used for the test (*Salmonella typhimurium TA 1535*) the umuC-gene is fused with a lacZ-gene. Therefore when the SOS-system is activated, β-galactosidase is produced, which can easily be detected by using a chromogenic substrate. Many substances are only genotoxic when metabolically activated by the biotransformation system. For this metabolic activation of progenotoxic substances a liver-S9-mix is added.

The microtiter plate assay is transformed into an automated on-line assay using suspended bacteria. A photometric detection similar to the microtiter

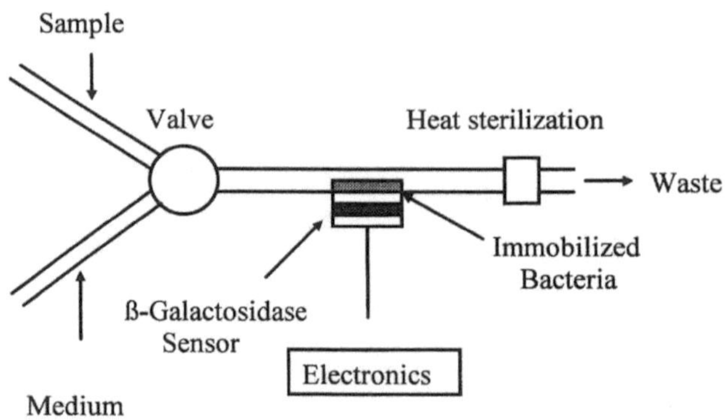

Figure 7. Flow scheme of the genotoxicity Sensor.

plate test is used. The working systems are validated by using model genotoxic substances such as 4-nitro-chinolin-N-oxide (4-NQO) and 2-aminoanthracene (2-AA). The flow scheme of the gentoxicity sensor is presented in Figure 7.

Immunotoxicity sensor

The resilience of a freshwater or marine organism is influenced by changes in the environment which are due to both natural and man-made pollution. By recording immunological resistance (phagocytosis) in terms of quality and quantity (Hansen et al., 1991; Hansen, 1992a), it is possible to detect pollution effects on organisms by an immunotoxicity sensor. Through phagocytic activity foreign particles and attached pollutants are digested (see Fig. 8).

The hemocytes of mussel feed on FITC (fluorescein isothiocyanate)-conjugated yeast cells or latex particles which are treated as foreign particles and attached pollutants (Hansen, 1992a; Krumbeck et al., 1994). The phagocytic activity is directly related to fluorescence and/or decrease in luminescence. Hemocytes play a major part in the immunological defence system of many invertebrates. Measurement of phagocytic activity offers ample opportunities for detecting unknown biotoxins by their influence on mussel immunology. A good example for the application of the phagocytic bioassay is the detection of algal toxins in marine and freshwater environments. The number of blooms of toxic algae has increased recently because of the eutrophication of freshwater and marine ecosystems. These blooms pose a threat not only to natural ecosystems and fisheries but also to human health. Since biotoxins excreted by algae have not been identified, a bioassay system such as phagocytosis or a biosensor for immunotoxicity could be extremely useful in detecting their effects.

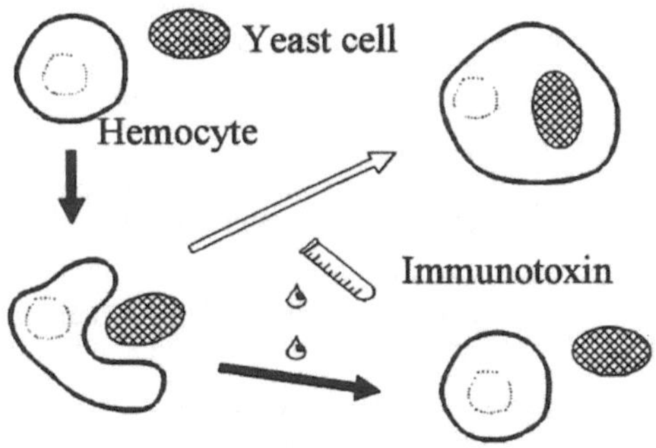

Figure 8. Principle of the immunotoxicity sensor.

Experiments of feeding mussels with toxic algae (*Chrysochromulina polylepis*) and non-toxic algae (*Isochrysis galbana*) and the resulting phagocytosis analysis have been reported by Krumbeck et al. (1994). For these experiments the phagocytosis was performed using the microplate technique. Developments in this direction are currently underway, the final measurements of the phagocytosis index using a fluorescence and/or luminescence microplate reader. This procedure has the potential for becoming an on-line immunotoxicity biosensor system.

Future prospects

Because there are large numbers of anthropogenic pollutants in excess of 100 000 in waterways, it is impossible to detect all these substances through chemical detectors. Presently there are several biomarkers that are effectively used in pollution monitoring and abatement programs. Principles associated with the different scales of biochemical processes relating to ecosystems are given in Figure 9 which shows the structural and functional hierarchies of biological responses and interactions as they relate to ecosystem complexity. To understand the complexity of the processes in the environment, we have to direct our efforts to the promotion of rapid and cost-effective testing and sensor methods. The use of on-line biosensors offers a good opportunity to evaluate "signals of early warning" and to understand in advance the changes in the environment.

The biosensor "signals" are helpful in promoting an environmentally sensitive and sustainable use of waterways and coastal zones for ecosystem health management. Biosensors generate continuous information. These "signals" in relation to time and space determine the ecosystem pulses describing disasters in large-scale systems caused by anthropogenic inputs. Such information would be highly helpful in proper understanding and managing of ecosystems and application of restoration measures and strategies. Biosensor detection approaches would be useful for generating in-

		Responses in biological systems
[min]		
10^6	(< 2 years)	*Ecosystem Level:*
		Alteration in ecosystems – redevelopment of
		the systems elements and structure.
	(0.5–1 year)	*Population Level:*
	(0.5–1 year	Population dynamics – self organization – reorganization
10^5	(1 – 12 months)	Change in growth and adaptation of the system.
	(20–120 days)	*Organismic Level (exposure):*
		– growth, reproduction, ELST (early life stage test), accumulation,
10^4		biotransformation: MFO, reaction with macromolecules,
		DNA-damage, repair, mutagenesis.
		Suborganismic Level:
	(1–3 days)	Symptoms in individuals: detoxification
		and regulation processes
10^3		– biochemical response ("MFO", ChE, phagocytosis)
		– change in behavior.
10^2		
10^1	(10 min)	*Early Warning Systems – On Line Monitoring "Biosensors"*
10^0		Input of pollutants (sublethal level)

Figure 9. Timescale responses of pollutants in biological systems: Biosensors to ecosystem levels. MFO = mixed function oxygenases, ChE = cholinesterase.

formation on early warning signals (initial onset) of environmental deterioration processes (Hansen, 1992b).

Progress has already been made using enzymes and bacterial bioluminescence detection systems. A future direction for effective biosensors at this time is the immunoassay technique (immunoreactors for screening of immunocompetence in different organisational levels of ecosystems). Genetically engineered bacteria (Côté et al., 1995) play an increasingly important role in the biosensor technique. One example is the umu-assay (DIN 38415-T3 and ISO/CD 13829) which is already part of the Federal Regulatory Programme after the Federal Water Act (§ 7a WHG) for effluent control. It can be transferred into an automatic genotoxicity sensor (Fig. 7) in order to monitor on-line genotoxicity using a standardized method.

The decision which has to be made is towards a summarizing effect parameter, like genotoxicity or toxicity in general, or a biosensor for the detection of single substances. For effluent and process control the measurement of sum parameters is more suitable. These sensors should be combined with specific sensors if there are special target compounds to be monitored.

For the interpretation of the alarm events detected by summarizing sensors instrumental chemical analysis (GC/MS, HPLC etc.) is indispensable. Even the specific sensor has to be calibrated by instrumental methods (e.g. GC/MS) because of cross reactivity. The missing link at this time is a robust biosensor in on-site operation, and there is a need for intercalibration studies with biosensors in different matrices (water, soil and air) in a day by day routine.

The future is likely to see sensors for detergents and endocrines. Especially the existing immunoassay based sensors have to be transferred into real matrices. In future there should be a data bank for antibodies, technique transfer and trainings as well as exchange of experiences with biosensors with real environmental matrices. The targets for on-line monitoring using biosensors for a better understanding of environmental processes are effluents, surface water, groundwater, deposit sites and remediation success control.

Once information on the toxic, genotoxic and immunotoxic potentials or endocrine effects in ecosystems are available by a battery of biosensors, effective water quality control measures can be easily applied for environmental management. Environmentally relevant "early warning signals" in ecosystems given by biosensors (Hansen, 1992b) would not only tell us the initial levels of damage and disasters, but could help to develop precautionary measures.

Acknowledgement
The development of the water-quality monitoring bacteria electrode has been supported under the Environment and Climate Programme of the European Commission (DG XII D1).

References

Côté, C., Blaise, C., Delisle, C. and Hansen, P.-D. (1995) A Miniaturized AMES Test Employing Bioluminescent Strains of *Salmonella typhimurium*. *Mutat. Res.* 345: 137–146.

Hansen, P.-D. (1992a) Phagocytosis in *Mytilus edulis*, a System for Understanding the Sublethal Effects of Anthropogenic Pollutants and the Use of AOX as an Integrating Parameter for the Study of Equilibria between Chlorinated Organics in *Dreissena polymorpha* Following Long Term Exposures. *In:* Neumann, D. and Jenner, H.A. (eds): *Limnologie aktuell*, Vol. 4 *The Zebra Mussel Dreissena polymorpha*. Gustav-Fischer-Verlag, Stuttgart, Jena, New York.

Hansen, P.-D. (1992b) On-Line Monitoring mit Biosensoren am Gewässer zur ereignisgesteuerten Probenahme. *Acta hydrochim. hydrobiol* 20(2):92–95.

Hansen, P.-D. (1995) The Potential and Limitations of New Technical Approaches to Ecotoxicology Monitoring. *In:* M. Richardson. (ed.) *Environmental Toxicology Assessment*. Taylor and Francis, London.

Hansen, P.-D. and Stein, P. (1994) Eucyano-Bacteria Electrode for On-Line Monitoring of Waterways for sampling Aimed at Relieving Disturbances. *In*: P. Bennetto and J. Büsing (eds) *Biosensors for Environmental Monitoring*. Technologies for Environmental Protection, Report 3, EUR 15622 EN, 14–19.

Hansen, P.-D., Pluta, H.-J. and Beeken, J.A. (1989) Biosensors for On-Line Monitoring of the Waterways and for Sample Taking Aimed at Relieving Disturbances, GBF Monogr., 1989, 13 (Biosensor); 113–116.

Hansen, P.-D., Bock, R. and Brauer, F. (1991) Investigations of Phagocytosis Concerning the Immunological Defence Mechanism of *Mytilus edulis* Using a Sublethal Luminescent Bacterial Assay (*Photobacterium phosphoreum*). *Comp. Biochem. Physiol* 100C(1/2): 129–132.

ISO (1996) Water quality – Determination of the Genotoxicity of Water and Waste Water using the umu-Test ISO/TC147/SC5/WG9 N10 (CD Draft January 1996).

Krumbeck, H. Elbrächter, M., Herbert, A. and Hansen, P.-D. (1994) Effects of algae toxins on the phagocytic activity of mussels hemocytes. *In*: *6th International Conference on Toxic Marine Phytoplankton*. Ed. Lassus. Nantes.

Meier, H., Lantreibecq, F. and Tranh Minh, C. (1992) Application and Automation of Flow Injection Analysis (FIA) Using Fast Responding Enzyme Glass Electrodes to Detect Penicillin in Fermentation Broth and Urea in Human Serum. *J. Autom. Chem.* 14(4): 137–143.

Oda, Y., Nakamura, S.I., Oki, I., Kato, T. and Shinagawa, H. (1985) Evaluation of the new system (umu-test) for the detection of environmental mutagens and carcinogens. *Mut. Res.* 147: 219–229.

Rao, S.S., Burnison, B.K., Efler, S., Wittekindt, E., Hansen, P.-D. and Rokosh, D.A. (1995) Assessment of Genotoxic Potential of Pulp Mill Effluent and an Effluent Fraction Using AMES-Mutagenicity and umuC-Genotoxicity Assays. *Environ. Toxic. Water Qual.* 10: 301–305.

v. Usedom, A. (1995) *Electroanalytical Investigations on Electrodes with Immobilized Organisms*. Master thesis, Berlin University of Technology.

Wittekindt, E., Werner, M., Reinicke, A. Herbert, A. and Hansen, P.-D. (1996) A Microtiter-Plate Urease Inhibition Assay – Sensitive, Rapid and Cost-Effective Screening for Mercury and other Heavy Metals in Water. *Environ. Technol.* 17: 597–603.

Frontiers in Biosensorics II
Practical Applications
ed. by. F.W. Scheller, F. Schubert and J. Fedrowitz

Biosensors for food analysis

A. Warsinke

Institute of Biochemistry and Molecular Physiology, University of Potsdam, c/o MDC Max-Delbrück-Center, D-13122 Berlin, Germany

Summary. Concerning speed, cost and on-line capabilities, biosensors offer attractive alternatives to existing methods for food analysis. They make monitoring and control of manufacturing processes possible. Furthermore, portable biosensors could be used for monitoring in manufacturing, retail and distribution of foods. An overview is given about existing biosensors for foodstuffs that could find applications in food industry.

Introduction

Since food and fodder products are complex mixtures of chemically diverse compounds, highly specific and reliable methods are needed for their analysis. Besides conventional chromatographic methods, e.g. HPLC or GC, enzymatic assays have found widespread application in food control (Bergmeyer and Graßl, 1983). The high specificity of the used enzymes allows the detection of one compound in the presence of a huge number of others. However, only for a few analytes do appropriate enzymes exist and for the determination of compounds at a low level (< µmol/l) enzymatic methods are only suitable if used in conjugation with enzymatic amplification methods, which are known to be reliable only under very strict operation conditions (Passonneau and Lowry, 1993). The application of new biorecognition elements could be a solution to these problems. Antibodies which have found widespread applications in immunoassays for the clinical area are potential candidates for the food area too. They offer high binding capabilities for their antigens and can be produced against nearly any substance, e.g. low-molecular weight pesticides, high-molecular proteins, and even microorganisms. In combination with enzymatic amplification systems enzyme immunoassays (EIA) allow detection limits at the subnano- to picomolar level. To date several commercial food immunoassay kits exist, and their number is on the increase. Application mainly focuses on the determination of food contaminants, e.g. pesticides, hormones, fungal and bacterial contaminations; however, natural food components, e.g. casein, soya, fish species and additives, e.g. flavours, sweeteners, antioxidants and gums (Allen and Smith, 1987; Lee and Morgan, 1993) are other target analytes.

Nevertheless, enzymatic assays and immunoassays are more or less time consuming, not suitable for on-line coupling, and require highly skilled labour.

Concerning improvements in speed, cost and on-line capabilities biosensors offer attractive alternatives to existing methods so that monitoring and control of manufacturing processes are becoming possible. Furthermore, portable biosensors could be used for monitoring in manufacturing, retail and distribution of foods. Outside the medical area, which is most important for biosensor development, the food industry has been considered to be the most important market for biosensors (*Biosensors: a new realism*, CBL Publications, 1991). But at present, as in other areas, the increasing number of publications and patents should not be used as a realistic indicator for the progress being made in creation of marketable products. Only a few types of biosensors (mostly for the determination of glucose, L-lactate, galactose and alcohol) have left the laboratory bench and are commercially availabe (Table 1). Because the food profit margins are low in comparison to those in the medical area, commercialisation of biosensors for food analysis will only be successful if the developed biosensors can fulfil a common need for analysis and find a gap between existing con-

Table 1. Selected companies involved in the development of biosensors for food analysis.

Company	Activity
ATI Orion, USA	Bio Selective Electrodes 5550 system: glucose, sucrose, lactose
Biometry, Germany	HPLC + biosensor, glucose, ethanol
Colera Messtechnik GmbH, Germany	on-line fermentation control: glucose, lactate, ethanol
Cranfield Institute of Technology, UK	Glucose, microbial contamination, methanol
Dosivit, France	MC2 Multisensor: glucose, sucrose, lactose, lactate, ethanol
Fuji Elictric Co., Japan	Gluco 20: glucose
Integrated Genetics, USA	DNA probes for detection of microbial contamination: (*Salmonella*)
Molecular Devices Corporation, USA	Threshold-System (based on light-addressable-potentiometric sensor): assay for DNA traces
NEC, Japan	FET biosensors: glucose, alcohol, L-lactate, glycerine
Oriental Electric Co., Japan	KV-101 freshness meter: degradation products of ATP
Pegasus Biotechnology, Canada	Microfresh: degradation products of ATP
Provesta Corporation, USA	Multipurpose Bioanalyzer: glucose, lactate, lactose, alcohol
Prüfgeräte-Werk Medingen GmbH	Industrial Module: glucose, L-lactate, lysine: lactose, glutamate, ascorbate in preparation
Setric Genie Industriel, France	Microzym-L: glucose, lactate, sucrose, lactose
Solea-Tacussel, France	Glucoprocesseur: glucose, L-lactate
Technicon Inc. Yellow Springs, USA	YSI 2000 automated analyzer/YSI 2700 Industrial analyzer: L-lactate, glucose, ethanol, methanol, sucrose, lactose, starch, galactose
TOA Electronics Ltd., Japan	Glu-11: glucose
Toyo Jozo, Japan	Biosensors for glucose, lactate, lipids
Universal Sensors, USA	Potentiometric electrodes: glucose, lactate, alcohol, lactose, cholesterol, various L-amino acids

ventional methods (for overviews see Luong et al., 1991; Roe, 1992; Icaza and Bilitewski, 1993; Griffith and Hall, 1993).

Before starting the development of a biosensor fulfilling industry requirements the following questions should be considered:

1. Where is the market?
2. Are there alternative methods, and what is their level of establishment?
3. What about the financial support for the biosensor development-profit margins of the users?
4. What about the sample matrix? Is the matrix changing?
5. Are there any temporary extreme conditions, e.g. high or low temperatures, gases, strong acids?
6. Which application: in-line, on-line, at-line, off-line, reusable or disposable?
7. Which measuring range, measuring time, shelf-life has to be covered?

Sugars

Most biosensors for food applications that have been described in the literature are biosensors for the determination of sugars. The reasons for this are both the widespread occurrence of sugars in all kinds of food products and the utilisation of the reliable and commercially available biosensor configurations using glucose oxidase as the biocomponent and an amperometric electrode as the transducer. This configuration has found widespread applications in the medical area, and most of the commercially available biosensors are based on this principle. The glucose oxidase reaction is followed by the measurement of oxygen consumption, hydrogen peroxide production or conversion of a mediator. In the food area this principle has been used for the determination of glucose by a variety of enzyme membrane sensors, flow-injection analysis systems (FIA) and screen-printed enzyme electrodes. In combination with other enzyme reactions in a linear or competition mode it allows the determination of other sugars, like disaccharides or polysaccharides (Table 1 and 2). A problem encountered with multienzyme sensors is the susceptibility to interferences caused by intermediates present in the food sample. Glucose present in a food sample will disturb the measurement of lactose, if the enzyme β-galactosidase is combined in a linear sequence with the enzyme glucose oxidase, and the presence of glucose will disturb the measurement of sucrose, if this is based on invertase combined with glucose oxidase. To solve this problem several approaches have been used in the past. One of them is the elimination of the intermediate with enzymatic methods before the sample is exposed to the biosensor. Unfortunately, this method is tedious and expensive. Another approach, which has been successfully applied by Scheller and Renneberg (1983) for the determination of sucrose

Table 2. Selected biosensors for the determination of sugars in foods (see also Wagner and Schmid, 1990).

Analyte	Enzyme, micro-organism	Detection principle	Life-time (d)	Linear range (mM)	Application	Ref.
glucose, sucrose	IN/GOD	amperom., O_2	20	0,2–2,8 0.5–7.0	instant cocoa	Scheller and Karsten, 1983
glucose, sucrose	fructosidase/ MUT/GDH	fluorim., NADH		0.01–03	fruit juices	Ogbomo et al., 1991 a
glucose	*Zymomonas*	pH-electrode		< 8 g/l	on-line monitoring	Park et al., 1995
fructose sucrose	*mobilis*/ IN			< 80 g/l < 60 g/l	(glucose production)	
glucose	GOD/POD	amperom., 1,1′-dimethyl-ferrocene		0.02–0.5	soft drinks	Yabuki and Mizutani, 1995
glucose	GOD	fluorim., O_2	16	0.1–500	wine, fruit juice	Dremel et al., 1989
glucose	GOD	amerom., 1,1′-dimethyl-ferrocene			molasses	Bradley et al., 1989
glucose	GOD	amperom., H_2O_2	120	20–200 mg/dl	honey, Cola, punch, sweet potato, apple juice	Wei et al., 1995
glucono-lactone	GDH/GOD	amperom., O_2	5	0.02–1	fermentation broth	Warsinke et al., 1991
maltose, sucrose,	GA/GOD IN/MUT/ GOD	amperom., O_2	7 28	0.03–2.5 0.1–6.0		Filipiak et al., 1996
lactose	*β*-Gal/GOD		7	1.0–6.0		
sucrose	IN/MUT/ polypyrrole/ GOD	amperom., H_2O_2, FIA	25	1.0–40.0		Schuhmann and Kittsteiner-Eberle, 1991
lactose	*β*-GAL/ GOD	amperom., H_2O_2	50	0.002–3.0	milk	Pfeiffer et al., 1990 a
lactose	*β*-GAL/ GOD	amperom., H_2O_2	28	0.01–100	milk	Yu et al.,1990
maltose	GA/GOD	amperom., H_2O_2, FIA		0.2–4	brewer's yeast fermentation	Varadi et al., 1993
pullulan	pullulanase/ GA/GDH	fluorim., NADH		2–20 mg/l		Ogbomo et al., 1991 b
xylose	*Glucono-bacter oxydans*		28	5.0–30		Reshetilov et al., 1996

GA, amyloglucosidase; *β*-Gal, *β*-galactosidase; GDH, glucose dehydrogenase; GOD, glucose oxidase; IN, invertase; MUT, mutarotase; POD, Horseradish Peroxidase.

in sugar beets and cocoa drinks, is the usage of an anti-interference layer in the form of immobilised glucose oxidase and catalase which is connected directly to the sucrose-converting enzyme sequence layer at the top of the electrode. When the glucose-containing sample diffuses through the anti-interference layer the glucose is converted to δ-gluconolactone and hydrogen peroxide, the latter being then eliminated by the catalase reaction. The signal by up to 2 mmol/l glucose has been eliminated in this way. The potential interference of glucose in sucrose measurements was tested with mixtures of 80 mmol/l sucrose and 5–50 mmol/l glucose. The sensor read-out for sucrose was not influenced significantly by the glucose concentration. The concentration interval studied exceeds the maximal glucose/sucrose ratio of most food samples. For the general use of an anti-interference layers in multienzyme sensors it has to be taken into account that the production of reproducible multienzyme membranes is still a problem. Another problem arises if the pH-optimum of the enzymes used in the anti-inference membrane is different from that of the enzymes used for the detection of the analyte. Normally, a compromise pH is used for measurement, which can influence the long-time stability or sensitivity of the biosensor. Besides biosensors based on the glucose oxidase reaction, several other tpyes of sugar-sensing biosensors have been developed that use oxidases, dehydrogenases and microorganisms in combination with different types of transducers, e.g. optical or potentiometric, and with special sampling systems, such as flow injection analysis (Chen and Karube, 1992) (Table 2).

Organic acids

Organic acids are natural constituents of several fruits (e.g. apple, strawberry, black and red currant, plum), honey, wine and vegetables (e.g. cauliflower, carrot, green kale, rhubarb, onion). Moreover, organic acids are used in food industry as aroma additives, antimicrobial agents and stabilisers. The organic acids determined most frequently with biosensors are amino acids, e.g. glutamic acid, glutamine, lysine (for overview see Pfeiffer et al., 1990b), fatty acids and hydroxy acids (Table 3). The general principles used in these biosensors are the same as for sugar determination.

The determination of citric and malic acid is of particular interest, because these compounds are the main acids in fruits and vegetables. Their concentrations reflect the total acid content of a food product. The determination is routinely performed by HPLC or enzymatic assays with photometric detection (Möllering and Gruber, 1966). Biosensor approaches for citrate determination have been described by Hasebe et al. (1990) and Hikima et al. (1992). In our group new multienzyme electrodes for citrate as well as malate have been developed which are based on an amperometric pyruvate oxidase electrode.

Table 3. Selected biosensors for the determination of organic acids in foods (see also Wagner and Schmid, 1990).

Analyte	Enzyme, micro-organism	Detection principle	Life-time (d)	Linear range (mM)	Application	Ref.
glutamate	GLOD	amperom., H_2O_2	10	0.001 – 1.0	seasonings	Wollenberger et al., 1989
glutamate	*E. coli*	potentiom. CO_2	21	0.8 – 6.0	glutamic acid fermentation broth	Karube and Sode, 1991
glutamine glutamate	glutaminase/ GLOD/Micro-peroxidase	chemi-. luminom. H_2O_2, FIA	70	0.05–5.0 0.025–1.0		Blankenstein et al., 1993
tryptophan	TMO	amperom., O_2	90	0.025 – 1.0	*E. coli* culture media	Simonian et al., 1995
lysine	LyOD/ Peroxidase	chemi-luminom. H_2O_2	60	0.01–1		Preuschoff et al., 1993
essential fatty acids	LOX	amperom., O_2	10	0.1 – 1.2	oils	Schoemaker and Spener, 1994
short-chain fatty acids	*Arthrobacter nicotiana*	amperom., O_2	7	0.11 – 1.7 butyric acid	milk	Ukeda et al., 1992
acetic acid	AK/PK/LDL	opt., FIA NADH	2	10 – 80	vinegar process control	Becker et al., 1993
lactate malate	LDH MDH	amperom., NADH	90	0.1 – 2 0.25 – 2	milk products	Silber et al., 1994
lactate malate	LDH/DI MDH/DI	amperom., O_2 mediator Vitamin K3, FIA	90	0.01 – 0.5 0.05 – 1.2	wines	Yoshioka et al., 1992
α-ketoiso-caproic acid, L-leucine	LeuDH	fluorim., NADH, FIA		20 – 100	*Corynebacterium glutamicum* fermentation	Kittsteiner-Eberle et al., 1989
acetic acid	*Trichosporon brassicae*	amperom. O_2	21	0.08 – 1.2	glutamic acid fermentation broth	Karube and Sode, 1991
citrate	CL/OAC/POP	amperom. H_2O_2, FIA	21	0.1 – 1.0	Grapefruit, Lemon	Matsumoto et al., 1995
citrate	CL/OAC/POP	amperom. O_2	18	0.001 – 1.0	various real samples	Gajovic et al., 1995
isocitrate	ICDH/ POD	amperom., O_2	1	0.1 – 2.0	fermentation broth	Schubert et al., 1985

DI, diaphorase; ICDH, isocitrate dehydrogenase; LDH, lactate dehydrogenase; MDH, malate dehydrogenase; AK, acetate kinase; GLOD, glutamate oxidase; LeuDH, L-leucine dehydrogenase; LOX, lipoxygenase; LyOD, L-lysine oxidase; OAC, oxaloacetate decarboxylase; PK, pyruvate kinase; POP, pyruvate oxidase; TMO, Tryptophan-2-monooxygenase.

The enzyme electrode for the determination of citrate uses a linear sequence of the following three enzymes (Gajovic et al., 1995):

$$\text{citrate} \xrightarrow{\textit{citrate lyase}} \text{oxaloacetate} + \text{acetate} \quad (1)$$

$$\text{oxaloacetate} \xrightarrow{\textit{oxaloacetate decarboxylase}} \text{pyruvate} + CO_2 \quad (2)$$

$$\text{pyruvate} + \text{phosphate} + O_2 \xrightarrow{\textit{pyruvate oxidase}} \text{acetylphosphate} + CO_2 + H_2O_2 \quad (3)$$

All three enzymes have been co-immobilised in a gelatine layer as described by Scheller et al. (1979). The optimal ratio of the three enzymes with respect to stability and performance was found to be 50 U citrate lyase, 20 U oxaloacteate decarboxylase and 20 U pyruvate oxidase per cm^2. The enzyme layer was sandwiched between two dialysis membranes and fixed to a modified Clark-type electrode. Hydrogen peroxide (+600 mV Ag/AgCl) as well as oxygen (−600 mV Ag/AgCl) indication was applicable, but with respect to interferences the oxygen indication has been preferred. The problem caused by varying oxygen content of the sample or buffer has been solved by use of a thermostated flow-through system. The resulting multienzyme electrode covers a linear measuring range up to 1 mmol/l citrate (Fig. 1) with a detection limit as low as 0.5 μmol/l and a response time of 5 min.

Since oxaloacetate and pyruvate are intermediates of the enzyme sequence, they both disturb the measurement of citrate. Calibration curves for both intermediates are also shown in Figure 1. As expected, the sensitivity drops from pyruvate to citrate. In real food samples only a high pyruvate content can cause a problem, because oxaloacetate is very unstable and decomposes to pyruvate. Several real samples have been assayed with the newly developed enzyme electrode. Good correlation to a reference method for citric acid was found (Table 4). Since the pyruvate concentra-

Table 4. Comparison of citric acid analysis using the developed biosensor and the enzymatic photometric reference method (Boehringer Mannheim).

Sample	Biosensor citric acid [mM]	Reference method citric acid [mM]
carrot juice (40%)	6.9	7.7 ± 5%
orange juice (100%)	45.0	47 ± 2%
apricot jam (40 g fruit/100 g)	71.0	69.0 ± 5%
tomato puree	92.0	87.9 ± 5%
fermentation medium (*C. purpurea*, early growth phase)	24.0	25.0 ± 5%
fermentation medium (*C. purpurea*, late-growth phase)	3.4	3.0 ± 5%

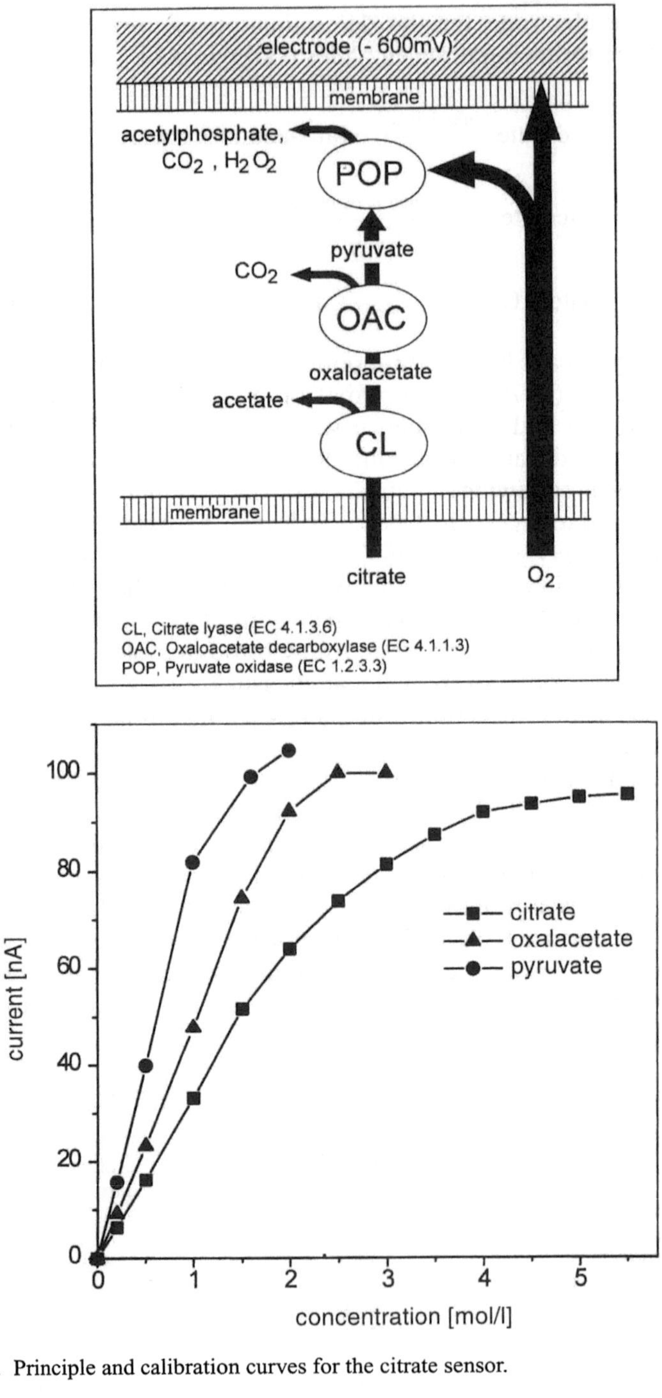

Figure 1. Principle and calibration curves for the citrate sensor.

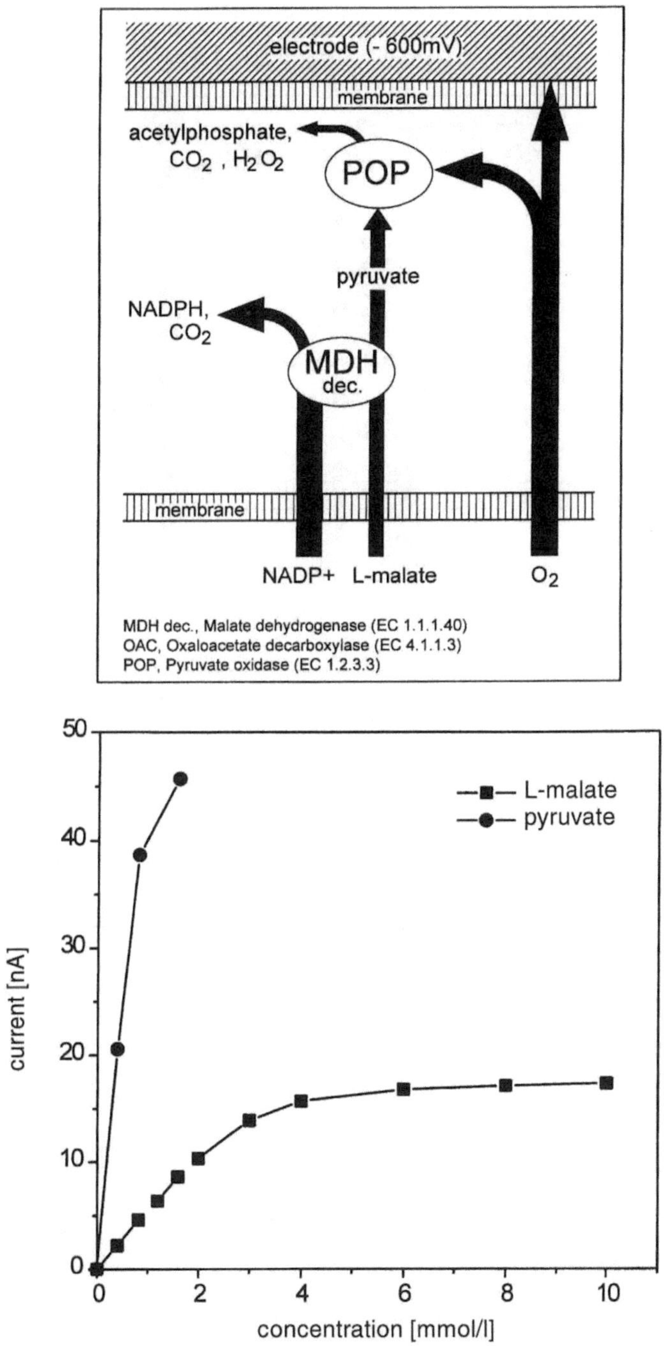

Figure 2. Principle and calibration curves for the malate sensor based on decarboxylating malate dehydrogenase.

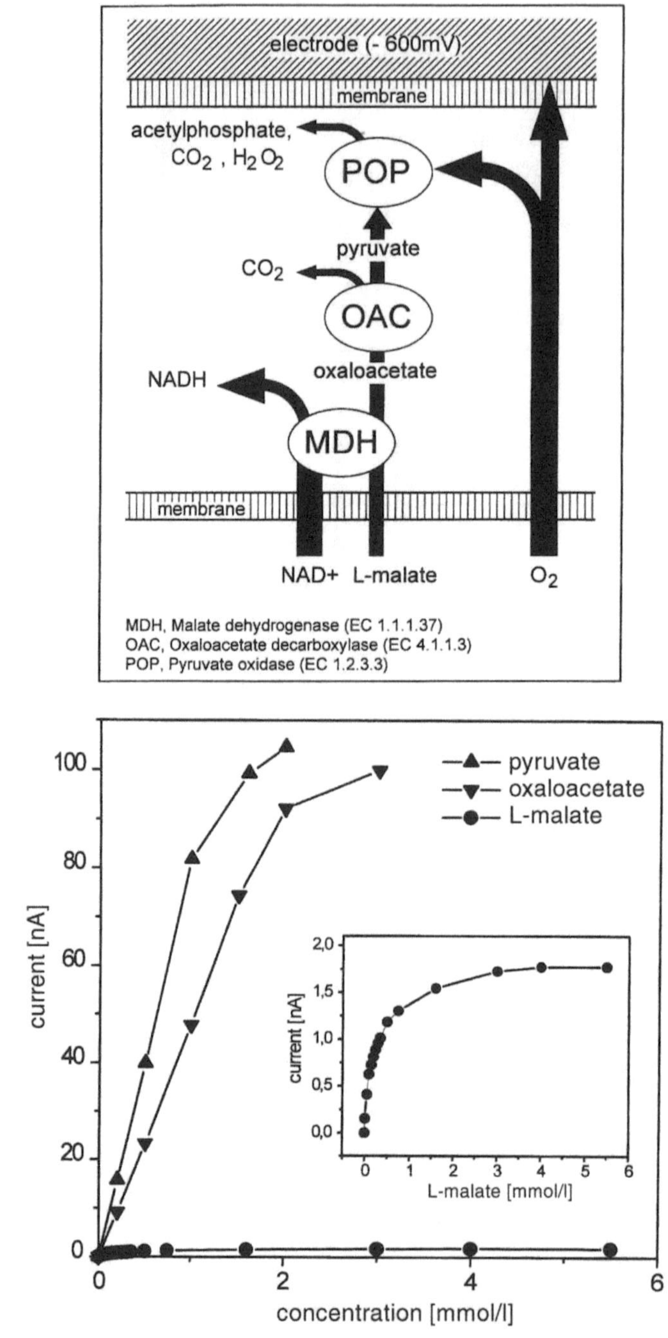

Figure 3. Principle and calibration curves for the malate sensor based on malate dehydrogenase and oxoloacetate decarboxylase.

tions of the investigated food samples were very low in comparison to the citrate levels, no special pretreatment was necessary. Citrate lyase is the most unstable enzyme in the sequence, because of deacetylation processes at the active site of the enzyme. To guarantee a functional stability of at least 5 days an enzyme loading of 50 U/cm^2 of citrate lyase within the membrane is required. More efforts are necessary to reach a higher stability, e. g. permanent regeneration of lost essential acetyl groups of the enzyme or the addition of stabilisers, as shown with comparable sensors (Gibson et al., 1992).

For the determination of malate the general biosensor configuration was the same as for citrate determination, but the enzymes were exchanged. Two different linear enzyme sequences were investigated and compared with respect to sensitivity as shown in Figure 2 and 3. The first enzyme sensor, which is based on the decarboxylating malate dehydrogenase (MDH dec.), covers a higher linear range than the second enzyme sensor and needs only two enzymes. Both enzyme electrodes are stable for at least 8 days. The enzyme sensor based on the MDH dec. is now under further investigation (for details see Gajovic et al., 1996). The next step in the direction to a total acid determination in foods will be the integration of both principles in one multienzyme sensor.

Other compounds

Besides sugars and organic acids, other substances in foods have to be quantified, e.g. pesticides, toxic compounds, mutagens, allergens, DNA-contaminations, vitamins and fat. Moreover, for transport and retail it is necessary to control food freshness. For the production of food products biosensors could find widespread application if they were able to control cleaning and disinfection processes with antimicrobial agents on-line.

The rapid total microbial detection (e.g. bacteria in dairy products) with biosensors is believed to be one of the most promising goals for biosensor research in the area of food analysis.

For the detection of microbial contamination of food samples enzyme sensors are not applicable. Immunosensors, which use antibodies as the biorecognition element, are more suitable because antibodies can be raised against surface antigens of various microorganisms in a very specific manner. In this way, an immunosensor can discriminate between different microorganisms. In combination with piezoelectric crystals as the transducer antibodies have been successfully used to detect *Candida albicans* (Muramatrsu et al., 1986), *Pseudomonas cepacia* (Nivens et al., 1993), *E. coli* and *Salmonella* (Prusak-Sochaczewski and Luong, 1990) in the range of 10^4–10^8 cells/ml. The same principle has been utilised for the detection of pesticides (e.g. atrazine, 2,4-D) in water (Guilbault et al., 1992; Minunni et al., 1994 a, b). For the detection of the pesticide parathion in the gas phase a piezoelectric immunosensor has been described by Ngeh-Ngwainbi et al. (1986). Nevertheless, for a continuous measurement of analyte concentrations, an immunosensor is not well suited because the equilibrium of the antigen-antibody reaction is far on the side of the antigen-antibody-complex. Harsh conditions (e. g. glycine/HCl buffer pH 2.0, 3 mol/l MgCl, 0.1 mol/l NaOH) have to be used to destroy this complex, which is necessary for the next measurement. These conditions will affect the binding capacity of the antibody and thus reduce the sensor sensitivity.

Another principle which should be mentioned in this context is the Polarisation Fluoroimmunoassay (PFIA). In contrast to the principle described above, which has a direct immunosensor configuration (no labelled compound is needed), it is based on the increase in the polarisation of the fluorescence of a small fluorescent-labelled antigen when bound by specific antibodies. If the sample contains the unlabelled analyte, it will compete with the fluorescent-labelled antigen for the binding sites of the antibody, and the polarisation signal will decrease. This principle has been used by Eremin et al. (1992) to detect sulphamethazine, chloramphenicol, benzylpenicillin and 2,4-D in buffer solutions, but has been considered to be applicable for food safety and quality assurance testing.

For control of food freshness efforts are underway to measure alcohol in fruits, aldehydes in fats, and histamine in fish with biosensors. Most biosensors in this area have been developed for fish freshness control, mainly via the determination of the K value in fish:

$$\text{K value (\%)} = \frac{\text{inosine} + \text{hypoxanthine}}{\text{ATP} + \text{ADP} + \text{AMP} + \text{IMP} + \text{inosine} + \text{hypoxanthine} + \text{uric acid}} \times 100$$

The K value is a parameter reflecting the decomposition of ATP, which starts after the death of the fish and is related to the spoiling process. If the K value is below 20%, it is possible to eat the fish (raw), if the K value is below 40% the fish can still be eaten after cooking, but if the K value is above 40% the fish should be avoided. The K value has been determined with various enzyme electrodes (Watanabe and Tanaka, 1991; Karube and Suzuki, 1992). Further possibilities to control fish freshness with biosensors are measurement of the produced biogenic amines, e.g. putrescine or cadaverine, during the spoiling process with enzyme electrodes (Chemnitus et al., 1992; Yang and Rechnitz, 1995) or the reduction of trimethylamine oxide (TMAO) to the trimethylamine (TMA) by spoilage bacteria with microbial sensors using *Penicillium decumbens* (Li et al., 1994) *or Pseudomonas aminovorans* (Gamati et al., 1991).

The determination of alcohol in food products is important for the control of brewing processes and quality of wine, beer and spirit. Most of the biosensors for alcohol determination are based on amperometric enzyme electrodes or fluorimetric methods, using the enzyme alcohol oxidase or alcohol dehydrogenase, respectively. Prinzing et al. (1990) and Ogbomo et al. (1993) have described a flow-through system with an integrated pervaporation module containing a membrane which is only permeable for volatile compounds. In this way interferences could be avoided, and in combination with an alcohol oxidase electrode ethanol concentrations of 5 to 100 mmol/l were determined. For the measurement of real samples (beers) the developed biosensor system correlated well with a standard test-kit. Karube and Sode (1991) have shown that it is also possible to use microorganisms for ethanol determination. The electrode, which was constructed by combination of *Trichosporon brassicae* and a Clark-oxygen electrode, was stable for at least 3 weeks and could determine ethanol in the concentration range of 0.043 to 0.49 mmol/l. The potential of the sensor was shown by measurement of yeast fermentation.

Glucosinolates, a group of the sulphur-containing heteroglycosides of the plant families *Brassicaceae, Resedacea, Caparidaceae and Tropeolaceae*, are undesired antinutrients producing goitrogenic and toxic effects in animals. Therefore the control of glucosinolates in food and fodder products originating from these plant families is of great importance. Conventional methods (e.g. GC) are time consuming, expensive and need educated staff. New enzyme sensors developed in the group of Macholán could be an alternative approach (Stancik et al., 1995) (Fig. 4, 5).

In a preincubation step the glucosinolates are digested to glucose and isothiocyanates by action of the enzyme myrosinase. One way to measure glucosinolates is to determine the release glucose by a glucose-sensing electrode (Koshy et al., 1988). Another way is to determine the release isothiocyanates through their inhibitory action on a tyrosinase electrode (Fig.4 A). Isothiocyanates are chemically converted into the corresponding thioureas which are strong inhibitors of the tyrosinase, as shown in Figure 5. For the measurement of the tyrosinase activity catechol is used as substrate. The associated oxygen consumption is determined with a Clark-type-electrode.

Not all glucosinolates can be detected by this method. After enzymatic digestion of the glucosinolate progoitrin, the corresponding isothiocyanate cyclises spontaneously and cannot be converted into the corresponding thiourea (Fig. 4B). In this way the content of progoitrin is not registrated by the tyrosinase sensor, but can be roughly estimated from the difference of the glucosinolate concentration determined by the glucose electrode and that determined by the tyrosinase electrode. The total amounts of glucosinolates found in processed samples of rapeseed meal agreed with those obtained using the reference gas chromatography (Stancik et al., 1995).

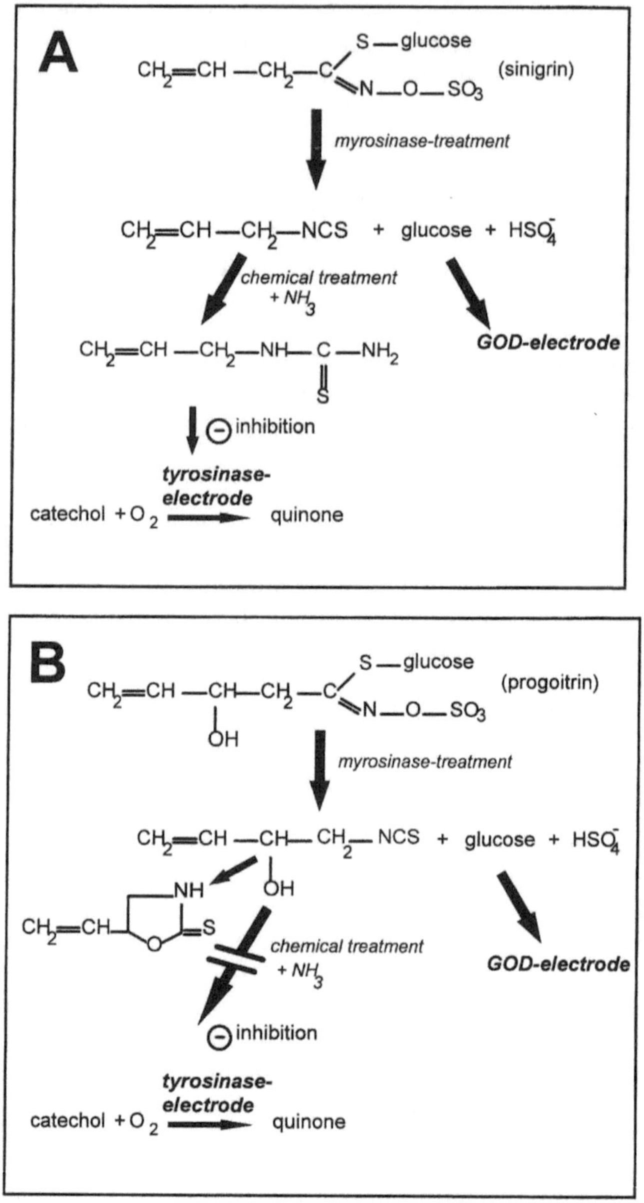

Figure 4. Principle of the determination of Sinigrin (A) and Progoitrin (B) with a tyrosinase and glucose electrode.

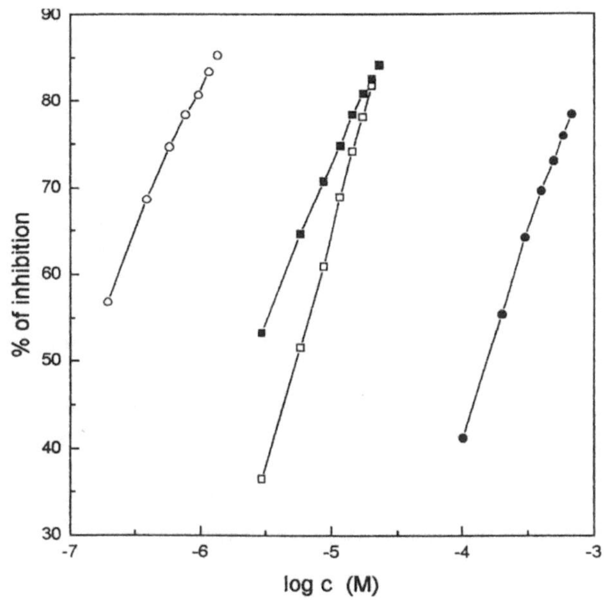

Figure 5. Inhibition of the tyrosinase electrode by phenylurea (○), allylthiourea (■), benzylthiourea (□) and phenylethylthiourea (●).

Conclusions

Although biosensors have been developed for a broad range of food analytes, such as sugars, amino acids, organic acids and proteins, only a few are really used in food industry. Some of the problems in the development of amperometric enzyme sensors as well as their possible solutions can be summarised as follows:

Problems	Possible solutions
Stability of the biocomponent	– suitable immobilisation method – addition of stabilisers – chemical modification – genetic engineering – use of synzymes or artifical biocomponents (e. g. imprinted polymers, abzymes, aptamers)
Need of coreactants	– coreactant recycling – direct e^--transfer – alternative biocomponent

Problems	Possible solutions
Interfering compounds	– oxygen indication – use of mediators – immobilisation matrix – anti-interference layer
Measuring range	– substrate recycling – additional diffusion barriers for substrate – alternative biocomponent

Thus, there is a repertoire of tools to overcome the problems often anticipated as hurdles in this field. This repertoire is broadened furthermore by using alternative biocomponents, such as antibodies or receptors, and alternative transducers, such as potentiometric, optical or thermal, so that for most applications suitable biosensors can be developed.

The most important problem seems to be the *way* of developing a marketable biosensor. The "trial and error" approach often used does not appear to be suited to generate the desired products. The adaptation of a biosensor developed for another field of application is equally ill-suited.

Another problem is the transfer of the biosensor production from the research laboratory bench to large-scale manufacture. The production at high quantities and low cost demands a high level of automation. Unfortunately, the production of biosensors often requires a high portion of manual processing, but with progress in transducer miniaturisation and protein engineering further impacts for mass production are expected in the future.

On the other hand mass production is only profitable if the developed biosensor fulfils a need for analysis. In food analysis this is expected for the determination of total sugar, bacteria, rancidity, alcohol, mycotoxins and pesticides.

In fact, the problem originating from the different constitutions of a myriad of food products are underestimated. Today, an ultimate biosensor that can measure bacteria in milk, butter or cheese seems to be unimaginable. As in many applications, the sample preparation is the rate-determining step rather than the biosensor response time. Thus, the author expects the potential of biosensors for food applications not in low-cost mass biosensors, which are produced in large-scale manufacture and have to compete with other products, but rather for some special applications where no extensive sample pretreatment is necessary. As in the medical field where new types of affinity sensors are likely to promote a breakthrough of biosensors, food analysis will benefit from this approach.

The food industry is known to be slow in applying new methods of analysis. But the influence of the consumers and new legislation will stimulate the search for modern test methods. If the industry defines clearly their needs and margins, researchers will find bioanalytical systems, including biosensors, which could really close the gap left open by conventional methods.

Acknowledgement
The author wishes to thank the Ministerium für Wissenschaft, Forschung und Kultur Brandenburg for the financial support of the Malate Project.

References

Allen, J.C. and Smith, C.J. (1987) Enzym-linked immunoassay kits for routine food analysis. *TIBTECH* 5:193–199.
Becker, T., Kittsteiner-Eberle, R., Luck, T. and Schmidt, H.-L. (1993) On-line determination of acetic acid in continuous production of *Acetobacter aceticus. J. Biotechnol.* 31:267–275.
Bergmeyer, H.U. and Graßl, M. (1983) *Methods of enzymatic analyis*, VCH, Weinheim.
Blankenstein, G., Preuschoff, F., Spohn, U., Mohr, K.-H. and Kula, M.-R. (1993) Determination of L-glutamate and L-glutamine by flow-injection analysis and chemiluminescence detection: comparison of an enzyme column and enzyme membrane sensor. *Anal. Chim. Acta* 271:231–237.
Bradley, J., Kidd, A.J., Anderson, P.A., Dear, A.M., Ashby, R.E. and Turner, A.P.F. (1989) Rapid determination of the glucose content of molasses using a biosensor. *Analyst* 114:375–379.
Chemnitius, G.C., Suzuki, M., Isobe, K., Kimura, J., Karube, I. and Schmid, R.D. (1992) Thin-film polyamine biosensor: substrate specificity and application of fish freshness determination. *Anal. Chim. Acta* 263:93–100.
Chen, C.Y. and Karube, I. (1992) Biosensors and flow injecton analysis. *Curr. Opin. Biotechnol.* 3:31–39.
Dremel, B.A.A., Schaffar, B.P.H. and Schmid, R.D. (1989) Determination of glucose in wine and fruit juice based on a fibre-optic glucose biosensor and flow-injection analyis. *Anal. Chim. Acta* 225:293–301.
Eremin, S.A., Landon, J., Smith, D.S. and Jackman, R. (1992) Polarisation fluoroimmunoassays for food contamination. *In*: M.R.A. Morgan, C.J. Smith and P.A. Williams (eds): *Food Safety and Quality Assurance: Applications of Immunoassay Systems.* Elsevier Applied Science, London and New York, pp 119–126.
Filipiak, M., Fludra, K. and Gosciminska, E. (1996) Enzymatic membranes for determination of some disaccharides by means of an oxygen electrode. *Biosens. Bioelectron.* 11(4): 355–364.
Gajovic, N., Warsinke, A. and Scheller, F.W. (1995) A novel multienzyme electrode for the determination of citrate. *J. Chem. Tech. Biotechnol.*63:337–344.
Gajovic, N., Warsinke, A. and Scheller, F.W. (1996) Comparison of two enzyme sequences for a novel L-malate biosensor. *J. Chem. Tech. Biotechnol.*, in press.
Gamati, S., Luong, J.H.T. and Mulchandani, A. (1991) A microbial biosensor for trimethylamine using Pseudomonas aminovorans cells. *Biosens. Bioelectron.* 6:125–131.
Gibson, T.D., Hulbert, J.N., Parker, S.M., Woodward, J.R. and Higgins, I.J. (1992) Extended shelf life of enzyme-based biosensors using a novel stabilization system. *Biosens. Bioelectron.* 7:701–708.
Griffith, D. and Hall, G. (1993) Biosensors-what real progress is being made. *TIBTECH* 11:122–130.
Guilbault, G., Hock, B. and Schmid, R. (1992) A piezoelectric immunobiosensor for atrazine in drinking water. *Biosens. Bioelectron.* 7:411–419.
Hasebe, K., Hikama, S. and Yoshida, H. (1990) Determination of citric acid by pulse polarography with immobilized enzymes. *Fresenius J. Anal. Chem.* 336:232–234.
Hikima, S., Hasebe, K. and Taga, M. (1992) New amperometric biosensor for citrate with mercury film electrode. *Electroanalysis* 4:801–803.
Icaza, M.A. and Bilitewski, U. (1993) Mass production of biosensors. *Anal. Chem.* 65: 525A–533A.

Karube, I. and Sode, K. (1991) Microbial sensors for process and environmental control. *In:* D.L. Wise (ed.): *Bioinstrumentation and biosensors*. Marcel Dekker, New York, pp 1–18.
Karube, I. and Suzuki, M. (1992) Microbiosensors for food analysis. *In*: P.R. Mathewson and J.W. Finley (eds): *Biosensor design and application.* American Chemical Society, pp 10–25.
Kittsteiner-Eberle, R., Ogbomo, I. and Schmidt, H.-L. (1989) Biosensing devices for the semiautomated control of dehydrogenase substrates in fermentations. *Biosensors* 4:75–85.
Koshy, A., Bennetto, H.P., Delaney, G.M., MacLeod, A.J., Mason, J.R., Stirling, J.L. and Thurston, C.F. (1988) An enzyme biosensor for rapid assay of glucosinolates. *Anal. Lett.* 21:2177–2194.
Lee, H.A. and Morgan, M.R.A. (1993) Food immunoassays: Application of polyclonal, monoclonal and recombinant antibodies. *Trends Food Sci. Technol.* 3:129–134.
Li, N., Endo, H., Hayashi, T., Fujii, T., Takai, R. and Watanabe, E. (1994) Development of a trimethylamine gas biosensor system. *Biosens. Bioelectron.* 9:593–599.
Luong, J.H.T., Groom, C.A. and Male, K.B. (1991) The potential role of biosensors in the food and drink industries. *Biosens. Bioelectron.* 6:547–554.
Matsumoto, K., Tsukatani, T. and Okajima, Y. (1995) Amperometric flow-injection determination of citric acid in food using free citrate lyase and coimmobilized oxalacetate decarboxylase and pyruvate oxidase. *Electroanalysis* 7(6):527–530.
Minunni, M., Skladal, P. and Mascini, M. (1994a) A piezoelectric crystal biosensor for detection of atrazine. *Life Chemistry Reports* 11:391–398.
Minunni, M., Skladal, P. and Macini, M. (1994b) A piezoelectric quartz crystal biosensor as a direct affinity sensor. *Anal. Lett.* 27:1475–1487.
Möllering, H. and Gruber, W. (1966) Determination of citrate with citrate lyase. *Anal. Biochem.* 17:369–76.
Muramatsu, H., Kajiwara, K., Tamiya, E. and Karube, I. (1986) Piezoelectric immuno sensor for the detection of *Candida albicans* microbes. *Anal. Chim. Acta* 188:257–261.
Ngeh-Ngwainbi, J., Foley, P.H., Kuan, S.S. and Guilbault, G.G (1986) Parathion antibodies on piezoelectric crystals. *J. Am. Chem. Soc.* 108:5444–5447.
Nivens, N.E., Chalmers, J.Q., Anderson, T.A. and White, D.T. (1993) Long-term, online monitoring of mirobial biofilms using a quartz crystal microbalance. *Anal. Chem.* 65:65.
Ogbomo, I., Kittsteiner-Eberle, R., Englbrecht, U., Prinzing, U., Danzer, J. and Schmidt, H.-L. (1991a) Flow-injection systems for the determination of oxidoreductase substrates: application in food quality control and process monitoring. *Anal. Chim. Acta* 249:137–143.
Ogbomo, I., Kittseinter-Eberle, R., Prinzing, U. and Schmidt, H.-L. (1991b) FIA-systems for the determination of maltose, lactate and volatile substances in beer, wine and fermentation broths. *In*: R.D. Schmid (ed.): *GBF-Monographs 14*. VCH, Weinheim, pp 209–216.
Ogbomo, I., Steffl, A., Schuhmann, W., Prinzing, U. and Schmidt, H.-L. (1993) On-line determination of ethanol in bioprocesses based on sample extraction by continuous pervaporation. *J. Biotechnol.* 31:317–325.
Park, J.K., Shin, M.C., Lee, S.G. and Kim, H.S. (1995) Flow injection analysis of glucose, fructose, and sucrose using a biosensor constructed with permeabilized Zymomonas Mobilis and invertase. *Biotechnol. Progr.* 11:58–63.
Passonneau, J.V. and Lowry, O.H. (1993) *Enzymatic analysis: a practical guide*. Humana Press, New Jersey, pp 85–110.
Pfeiffer, D., Ralis, E.V., Makower, A. and Scheller, F. (1990a) Amperometric bienzyme based biosensor for the detection of lactose – Characterization and application. *J. Chem. Tech. Biotechnol.* 49:255–265.
Pfeiffer, D., Wollenberger, U., Makower, A., Scheler, F., Risinger, L. and Johansson, G. (1990b) Amperometric amino acid electrodes. *Electroanalysis* 2:517–523.
Preuschoff, F., Spohn, U., Weber, E., Unverhau, K. and Mohr, K.-H. (1993) Chemiluminometric L-lysine determination with immobilized lysine oxidase by flow-injection analysis. *Anal. Chim. Acta* 280:185–189.
Prinzing, U., Ogbomo, I., Lehn, C. and Schmidt, H.-L. (1990) Fermentation control with biosensors in flow-injection systems – Problems and progress. *Sensor. Actuator. B*1:542.
Prusak-Sochaczewski, E. and Luong, J. (1990) A new approach to the development of a reusable piezoelectric crystal biosensor. *Anal. Lett.* 23:401–409.
Reshetilov, A.N., Donova, M.V., Dovbnya, D.V., Boronin, A.M., Leathers, T.D. and Greene, R.V. (1996) FET-micobial sensor for xylose detection based on Gluconobacter oxydans cells. *Biosens. Bioelectron.* 11:401–408.

Roe, J.N. (1992) Biosensor research. *Pharmaceut. Res.* 9(7):835–844.
Scheller, F. and Karsten, C. (1983) A combination of invertase reactor and glucose-oxidase electrode for the successive determination of glucose and sucrose. *Anal. Chim. Acta* 155: 29–36.
Scheller, F. and Renneberg, R. (1983) Glucose-eliminating enzyme electrode for direct sucrose determination in glucose-containing samples. *Anal. Chim. Acta* 152:265–269.
Scheller, F., Pfeiffer, D., Jänchen, M., Seyer, I., Siepe, M. and Pittelkow, R. (1979) *GDR Patent 127843*.
Schoemaker, M. and Spener, F. (1994) Enzymatic flow-injection-analysis for essential fatty acids. *Sensor. Actuator.* B19:607–609.
Schubert, F., Kirstein, D., Abraham, M., Scheller, F. and Boross, L. (1985) Horseradish peroxidase based bioenzyme electrode for isocitrate. *Acta Biotechnol.* 4:375–378.
Schuhmann, W. and Kittsteiner-Eberle, R. (1991) Evaluation of polypyrrole/glucose oxidase electrodes in flow-injection systems for sucrose determination. *Biosens. Bioelectron.* 6:263–273.
Silber, A., Bräuchle, C. and Hampp, N. (1994) Dehydrogenase-based thick-film biosensor for lactate and malate. *Sensor. Actuator.* B18–19:235–239.
Simonian, A.L., Rainina, E.I., Wild, J. and Fitzpatrick, P.F. (1995) A biosensor for L-tryptophan determination based on recombinant *Pseudomonas savastanoi* tryptophan-2-monooxygenase. *Anal. Lett.* 28:1751–1761.
Stancik, L., Macholan, L., Pluhacek, I. and Scheller, F. (1995) Biosensing for rapeseed glucosinulates using amperometric enzyme electrodes based on membrane bound glucose oxidase or tyrosinase. *Electroanalysis* 7(8):726–730.
Ukeda, H., Wagner, G., Biletewski, U. and Schmid, R.D. (1992) Flow injection analysis of short-chain fatty acids in milk based on a microbial electrode. *J. Agric. Food Chem.* 40:2324–2327.
Varadi, M., Adanyi, N., Nagy, G. and Rezessy-Szabo, J. (1993) Studying the bienzyme reaction with amperometric detection for measuring maltose. *Biosens. Bioelectron.* 8:339–345.
Wagner, G. and Schmid, R.D. (1990) Biosensors for food analysis. *Food Biotechnol.* 4(1): 215–240.
Warsinke, A., Renneberg, R. and Scheller, F. (1991) Amperometric multienzyme sensor for determination of D-glucono-δ-lactone. *Anal. Lett.* 24:1363–1373.
Watanabe, E. and Tanaka, M. (1991) Determination of fish freshness with a biosensor system. *In*: D.L. Wise (ed.): *Bioinstrumentation and biosensors*. Marcel Dekker, New York, pp 39–73.
Wei, D., Lubrano, G.J. and Guilbault, G.G. (1995) Dextrose sensor in food analysis. *Anal. Lett.* 28(7):1173–1180.
Wollenberger, U., Scheller, F., Böhmer, A., Passarge, M. and Müller, H.-G. (1989) A specific enzyme electrode for L-glutamate-development and application. *Biosensors* 4:381–391.
Xu, Y., Guilbault, G.G. and Kuan, S.S. (1990) Fast responding lactose enzyme electrode. *Ezyme Microb. Technol.* 12:104–108.
Yabuki, S. and Mizutani, F. (1995) Modifications to a carbon paste glucose-sensing enzyme electrode and a reduction in the electrochemical interference from L-ascorbate. *Biosens. Bioelectron.* 10:353–358.
Yang, X. and Rechnitz, G.A. (1995) Dual enzyme amperometric biosensor for putrescine with interference suppression. *Electroanalysis* 7(2):105–108.
Yoshioka, S., Ukeda, H., Matsumoto, K. and Osajima, Y. (1992) Simultaneous flow injection analysis of L-lactate and L-malate in wine based on the use of enzyme reactors. *Electroanalysis* 4:545–548.

Frontiers in Biosensorics II
Practical Applications
ed. by. F.W. Scheller, F. Schubert and J. Fedrowitz

Challenges in the development of (bio)chemical sensors for whole blood medical diagnostic applications

H. Luedi

Ciba Corning Diagnostics, Medfield MA, 02052, USA

Summary. The following article is based on experience in research and development of (bio)chemical sensors for medical diagnostic applications. It is not known to the author if the ideas presented below are applicable in other areas of sensor R & D.

Introduction

The most famous example of a (bio)chemical sensor is a planar, miniaturized (electrochemical) glucose sensor based on Glucose Oxidase proposed to be used in the diagnosis, treatment and monitoring of diabetic patients.

Despite a significant number of publications on glucose sensors and electrodes (Fig. 1), most of the tests are still done using technologies developed during the 1970's (Fig. 2). There is only one company (Medisense) with a

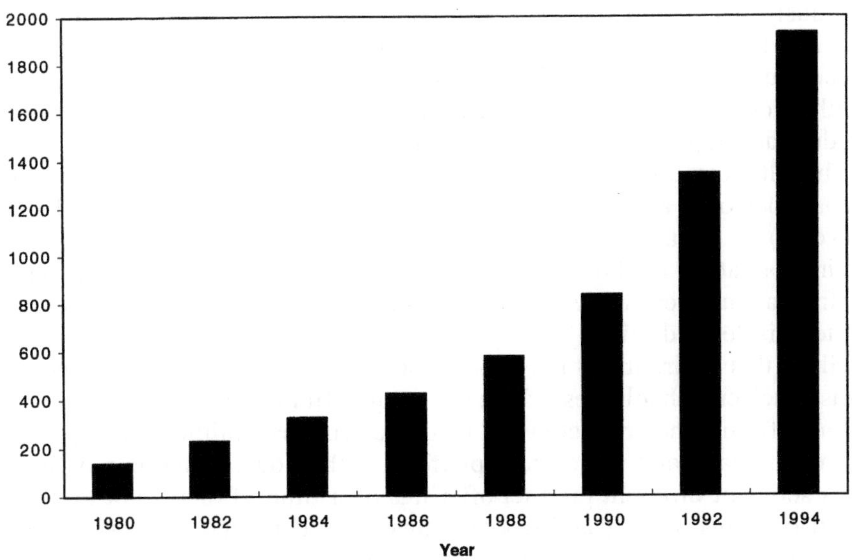

Figure 1. Number of publications on glucose electrodes and glucose sensors from 1980 to 1994.

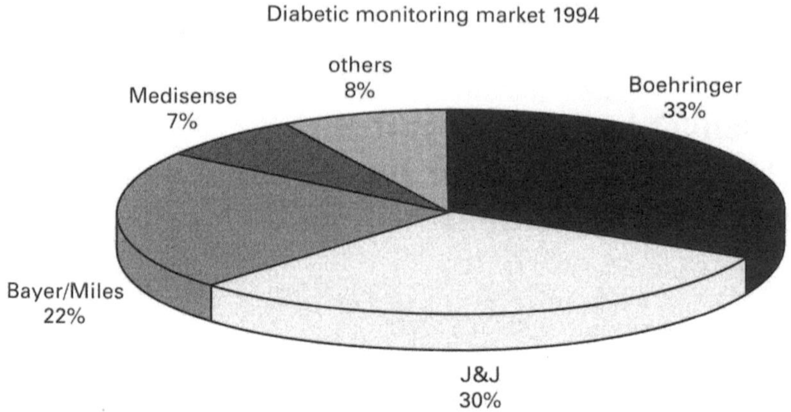

Figure 2. Relative market share of the top four companies in the diabetic monitoring market in 1994. (Total market: 1.3 billion $). Only one (Medisense) uses biosensor technologies.

significant market share selling glucose sensors based on technology developed in the 80's and 90's.

This was confirmed at the recent Transducers '95/Eurosensor IX conference in Stockholm (June 25–29, 1995), where industrial speakers unanimously agreed that for new sensors and sensor technologies it takes 8–12 years from the basic research idea to the break-even point (see Peterson et al., 1995).

Researchers at universities and those at corporate research centers are obviously disappointed by the facts and figures presented so far. This article will focus on some of the elements necessary to shorten time to successful marketing. It will propose research guidelines which significantly increase the acceptance of basic and fundamental research by the product development departments in industrial organizations.

The ultimate goal of the commercialization has to be satisfaction of a customer over an extended period of time (normally more than 10 to 25 years) for a reasonable profit. To assure this, most companies develop a business strategy ("who" are we and "who" is our customer), which is redefined about every three years. From that strategy an R & D/technology strategy is derived (Fig. 3).

Given this environment the most difficult task R & D managers encounter is to select technologies, which do not only fit into the business strategy, but also fit into the producton and manufacturing capabilities and expertise of the company and represent a platform technology, which can be used for a series of different product offerings over time. If this is not the case significant investments and risks are involved. This type of consideration often becomes the determining factor in the justification for the implementation of a new technology.

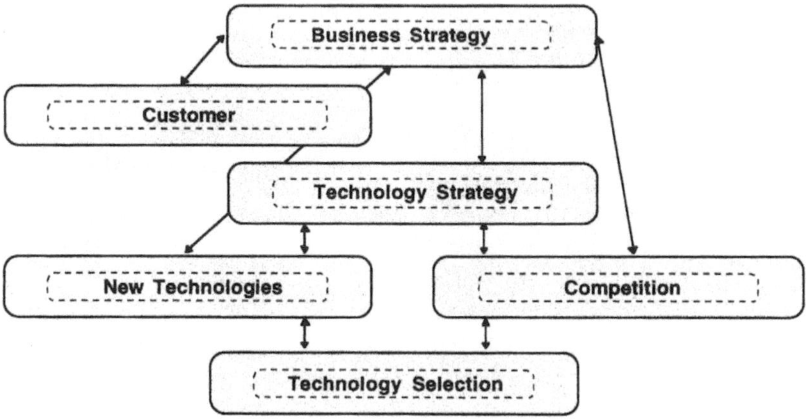

Figure 3. The business/technology environment of a medical diagnostic company.

For instance it is obvious that having production processes for an enzyme-based glucose sensor and a completely different one for a lactate sensor can never be justified in a manufacturing environment. Further, much money and effort would have been wasted in the R & D of glucose sensors if a non-invasive method became available today. Very recently promising results and commercially available instruments were announced by Duncan (1995) and Diasence Inc. (1995) Again, this illustrates the criticality of "Technology Selection" for a company.

The research guidelines

(Please note, this list is not complete. Add (delete) at your own risk.)

1 *Know your "customers"*

Do not copy the justification for your research from previous publications. What makes you believe that they are right? If you (or your "professor") would like to contribute to the field of medical diagnostics, call the clinical lab director of your local hospital and ask him for a tour. Prepare a few questions. Maybe you will be allowed to talk to a nurse or a physician. If your work is sponsored by a medical diagnostic company, ask to do the visit with the local sales representative. I am not aware of any situation where this could not be realized. Even though this will give you "only" one opinion, asking about how other hospitals are doing the same task or if this is a more of a general rather than specific procedure, together with your scientific "instinct" and common sense will allow you to draw meaningful conclusions. And, do not "believe" all the bar and pie graphs in market research studies. Somebody with insufficient knowledge and no experience about the market will draw the wrong conclusion, guaranteed.

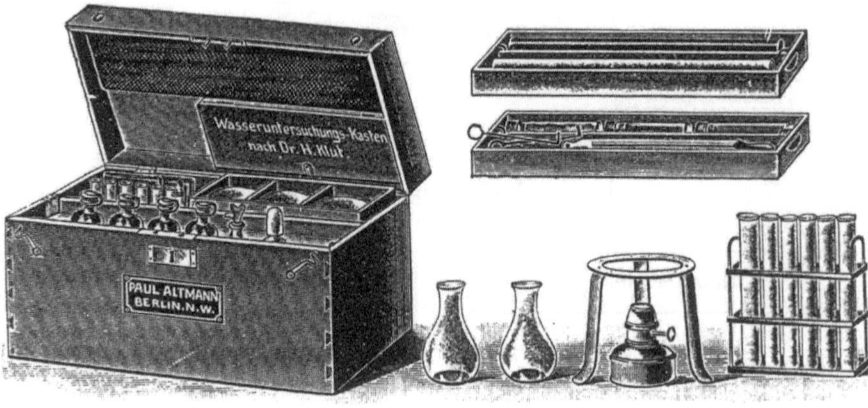

Figure 4. Portable kit for the investigation of water accoring to Klut. Original price 40 DM. Reproduced from: Untersuchung des Wassers an Ort und Stelle, Dr. Hartwig Klut, 3rd edition, 1916, Julius Springer Publishers.

2 *State-of-Art*

Before starting a (bio)chemical sensor research project, be aware of the possibility of completely different technologies that are commonly used to complete a task, test or service. How easy is it to use the current method, how much does it cost, how fast can it be done, who is doing it, and most importantly, why it is done and the manner in which it is done are questions to be answered very thoroughly. Know the literature, read and *understand* it. Do not limit your research to the last few years. A lot of good research has been done over the past *one hundred* years. For instance a portable, small analytical system for "on the spot" environmental control was proposed long before we cared about miniaturization (Fig. 4).

3 *Systems*

Any sensor, when becoming a part of a commercial product will become a part of a system. The "system" will constrain the flexibility in the "design" of the sensor. A glucose sensor e.g. integrated into a bench top blood gas/blood electrolyte instrument has not only to fit into "form and function", but also has to be calibrated with (existing) on board calibrants, as well as (existing) Quality Control (QC) materials.

The relative costs for the development of a glucose sensor for the Ciba Corning blood gas/blood electrolyte analyzers are given in Figure 5.

Expanding this thought, other parts of the system need consideration: Mechanical and electrical engineering, fluidics, software, information management, industrial design, software, software quality assurance, customer

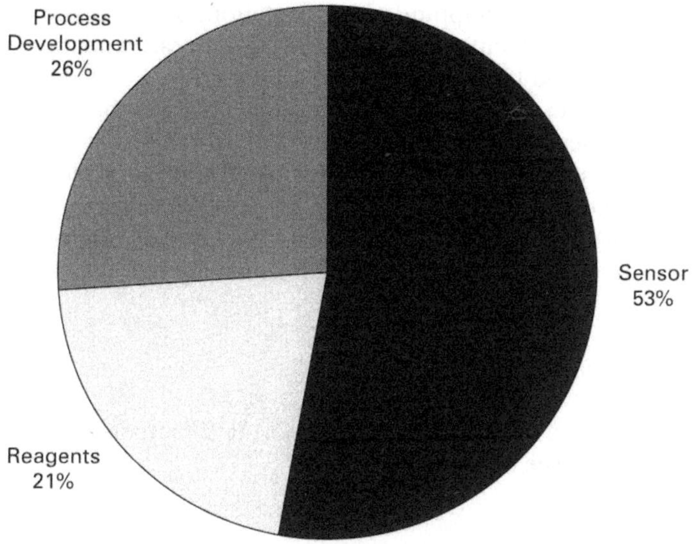

Figure 5. Relative cost of Sensor Development for a glucose sensor.

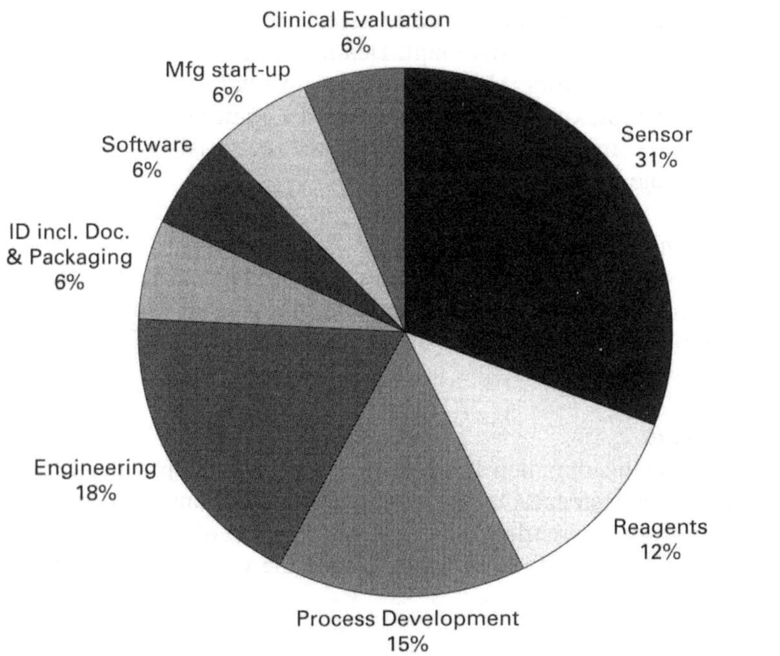

Figure 6. Relative cost for a glucose sensor (as in Fig. 5) including system integration and launch.

service, packaging and shipping, documentation, manuals and last but not least process development and manufacturing process validation to comply with FDA, GMP and ISO standards.

Integrating of sensors into new or existing systems, therefore, becomes a formidable research *and* development task, which you only undertake if you have a reasonably high probability of (financial) success. Figure 6 again shows the distributon of R & D spending during the integration of a glucose sensor into the Ciba Corning Blood Gas, Blood Electrolyte Analyses. This time the complete system is considered.

4 *Useful experiments*

The following are a few hints how to support R & D managers in their technology selection (see above).

4.1 *Statistics*

Accuracy, precision, bias to reference and maybe even the concept of total error should be known, understood and *applied* in every publication. Compare your numbers to the currently used technology (see above). Your performance will *not* improve significantly during development and commercialization. Factors which contribute to total error are increasing with the added complexity of every step toward a commercial system. Also, I have yet to find somebody who, over time, will be satisfied with less than currently available performance.

4.2 *Calibration*

The simple generation of a calibration curve should never be the reason for a publication or a Ph.D. thesis. A limited amount of "real" and "unknown" samples, as well as spiked samples, are needed to make a useful assessment of your technology possible.

4.3 *Interferences*

Researchers specialized in their field will be able to come up with a list of the most probable interferences using existing lists (National Committee for Clinical Laboratory Standards 1986). Test for the effect of these interferences. This does not take a lot of time and adds a lot to the value of your research.

4.4 *Manufacturing*

Carefully select the methods/processes used in the construction of the (bio)chemical sensor. Make sure that you can repeat these methods and proceses at least several times. To get to the statistics described in 4.1 will force you to repeat most of the experiments several times. How else can you be certain that you have observed a reproducible and real phenomenon?

4.5 *Use-life/Shelf-life*

Use-life, shelf-life and shipping studies can only be conducted in the "final" packaging and with "final" system. Preliminary indications of potential problems with initial designs are necessary to have sufficient confidence to proceed with research and development.

5 *The difference between reseach and development or: The chalenge of development*

Lastly, I would like to comment on the difference between research and development. In development the degree of freedom is greatly reduced. Many avenues are closed by system design factors and limitations. I firmly believe that development is as challenging as research. Since you are dealing with a *system*, creativity, excellent science, new ideas, challenging problems, detailed and profound knowledge and understanding of your field, and the requirement for interaction with engineers, marketing, service and others will train and educate you. Your horizon will broaden and you will be significantly challenged (sometimes to the extreme). But nothing is more rewarding in the career of an R & D scientist than releasing a product to the market and being complemented by customers for its performance.

Development really is fun!

To support this statement the successful development of a glucose- and a lactate sensor, integrated into a "conventional" blood gas/blood electrolyte analyzer should be mentioned (Ciba Corning's 860 system). Innovative approaches to development challenges such as whole blood measurement, large dynamic range, extended use- and shelf-life for sensors and calibration solutions, constant polarization at 37°C, interference correction using a "corrective" electrode, ratio of aqueous to whole blood measurement, and correlation (bias) to reference methods were necessary to realize a product which was highly reliable, accepted and (hopefully) appreciated by customers. (See Maley and D'Orazio (1995), McCaffrey (1995) and D'Orazio and Parker (1995) for details on reference methods, cover membrane development and interference corrections).

Acknowledgements
I would like to thank Richard Mason for carefully reading the manuscript.

References

Diasense Inc. (1995) has begun marketing its non-invasive blood gas sensor. *Sensor Business News* Sept., 1995

D'Orazio, P. and Parker, B. (1995) Interference by the Oxidizable Pharmaceuticals Acetaminophen and Dopamine at Electrochemical Biosensors for Blood Glucose. *Clin. Chem.* 41(6): 156.

Duncan, A., Hannigan, J., Freeborn, S.S., Rae, P.W.H., McIve, B., Greig, F., Jonston, E.M., Binnie, D.T. and MacKenzie, H.A. (1995) A Portable Non-Invasive Blood Glucose Monitor. *Transducer* '95, *Abstract 348-B13*, Stockholm, Sweden.

Maley, T.C. and D'Orazio, P. (1995) Biosensors for Blood Glucose: A New Question of What is Measured and What Should by Reported. *Clinical Laboratory News* 21(1): 12–13.

McCaffrey, R.R., D'Orazio, P., Mason, R.W., Maley, T.C. and Edelman, P.G. (1995) Clinically Useful Biosensor Membrane. *In*: D.A. Butterfield (ed.): *Development Biofunctional Membranes*. Plenum Publishing Corp., in press.

National Committee for Clinical Laboratory Standards (1986) Interference testing in clinical chemistry; Proposed Guideline. *NCCLS publication EP7-P*. Villanova, Pa.: NCCLS; (NCCLS Vol. 6 No. 13).

Peterson, K., (Lucas Novasensors, USA), Igarashi, I., (Toyota Physical and Chemical Res. Inst., Japan), Andren, G.L. (Pharmacia Biosensors AB Sweden), Horntvedt, S. (Sensonor, Norway) and Meixner, H. (Siemens AG, Germany) (1995) Session: Industrialization of Sensors. *Transducer '95* Stockholm, Sweden.

Frontiers in Biosensorics II
Practical Applications
ed. by. F.W. Scheller, F. Schubert and J. Fedrowitz

Commercial biosensors for medical application

D. Pfeiffer

BST Bio Sensor Technologie GmbH, D-13156 Berlin, Germany

Summary. Biosensors today are represented on the market worldwide by an increasing number of enzyme electrodes working in various areas of medical diagnostics, and by a few opto-immuno-sensor-based analytical systems suited for protein research in pharmaceutical chemistry. Enzyme electrodes for metabolites and enzyme activities have been commercially available for about 20 years. Such sensors are successfully applied in laboratory autoanalyzers, point-of-care-systems, patient self-monitoring disposable probes, intensive care analyzers, and for on-line monitoring of diabetics.

Introduction

To establish any biochemical sensor on the market it is essential to meet the necessities of the market as well as the requirements and expectations of the potential user. To meet these expectations the sensor has to be given optimal biochemical, technical and economical characteristics.

Presently there are various deficiencies in medical diagnosis that force the development of new analytical systems. Whereas it is sufficient to determine substances like urea, creatinine, sodium and cholesterol within periods of hours or even days (Mascini and Vadgama, 1993), the accurate monitoring of biochemical parameters like neutrotransmitters, lactate, potassium, oxygen, CO_2, NO, and other free radicals demands measurements within a few minutes. Thus, the precise identification and monitoring of short-lived substances and rapidly changing concentrations of other metabolites is possile only by high sampling frequencies or, ideally, real-time monitoring.

The demand for an improved efficiency of tumor and HIV diagnostics and control of therapies, and the realization of a successful point-of-care diagnosis with sufficient accuracy are further driving forces for the development of novel analytical systems. All these fields are challenges to biosensor development and define an enormous potential market that has not been covered by commercial systems up to now.

The main reasons for the limited market success of biosensors are the limited functional stability being in the range of months rather than years – which is desired by many users who are pampered by pH or pacemaker electrodes – and a drift behavior during real time analysis. However, one has to keep in mind that the analytical field dealing with biosensors is just 35 years 'old' as compared to more than a hundred years of classical analytical chemistry. Obviously, not all open analytical problems may be solved by analytical systems working with biological components. However, it is possible to contribute significantly by appropriate development and application of biospecific sensors.

Biochemical design of biospecific electrodes

Appropriate measuring range, specificity for the target analyte, sensitivity, measuring time and stability are the characteristics of enzyme electrodes that should be optimized with respect to biochemistry and its adequate combination with an appropriate transducer and analyzer concept. Normally a compromise has to be accepted with respect to priorities of the field of application.

The latter will dictate the *measuring range* of an enzyme electrode. The signal-concentration dependence for electrochemical biosensors is linear between one and three concentration decades. Usually the lower limit of detection is at 200 μmol/l with potentiometric and 1 μmol/l with amperometric enzyme electrodes. This allows the measurement of metabolites to be performed in *prediluted* whole blood, plasma, or urine samples. Such sensors are applied in commercial autoanalyzers.

A further decrease in the lower detection limit to the nanomolar range has been demonstrated with several substrate recycling systems based on coupled enzymes (Schubert et al., 1985; Yang et al., 1991; Wollenberger et al., 1993; Scheller et al., 1995), cyclic enzyme-chemical reactions (Hasebe et al., 1994), or combinations of enzymes and electrochemical reactions (Wasa et al., 1984; Mitzutami et al., 1991). In this way, the measurement of substances occurring in the lower nano- or picomolar range has been realized, e.g., dopamine, epinephrine, and norepinephrine. Up to now all these sensors have not worked in commercial instruments because of their limited reproducibility and stability. Significant improvement of these characteristics may be expected in the next years.

To realize the analysis of metabolites in undiluted whole blood, plasma, and urine without any preanalytics the measuring range of the sensor has to be shifted upwards by one or two orders of magnitude. This has been achieved by the inclusion of diffusion barriers between the sample and the enzyme layer (Scheller et al., 1978; Pfeiffer et al., 1992). Based on these enzyme membranes lactate and glucose measuring portable devices have been commercialized (BIOSEN/EKF Magdeburg, Germany).

The *response time* of enzyme electrodes is influenced by the thickness of the membranes and their diffusion resistance of the analyte. Whereas the response time of potentiometric enzyme electrodes averages 2–10 min, with amperometric sensors an assay can be conducted within a few seconds up to one minute. This permits up to several hundred determinations per hour depending on the peripheric flow management. The application of a glucose oxidase membrane with a characteristic diffusion time of 24 s allows 60 samples per hour to be measured in a stirred measuring cell (Pfeiffer et al., 1989). The integration of this membrane into an optimized air-segmented flow system with a measuring chamber of less than 10 µl results in a measuring frequency of 120/h (Scheller et al., 1989). This has been further improved to 300/h (Olsson et al., 1986) by use of a non-segmented flow system with 1.5 µl sample volume.

Increasing complexity of the biochemical reaction system, e. g., by coupled enzyme reactions or additional diffusion limitations may bring about an increase in the overall measuring time. Thus, the addition of an appropriate polycarbonate membrane to the glucose oxidase layer, while resulting in a tenfold extension of the linear range, causes an increase of the characteristic diffusion time to 50 s.

The *specificity* of enzyme electrodes is determined by both the respective enzyme and contributions of electrochemical interferences. Several possibilities have been studied to decrease the influence of electrochemically interfering substances: e. g. the application of an oxidation potential as low as possible, the use of optimal immobilization matrices and cross-linkers, the arrangement of an anti-interference layer in front of the enzyme electrode, and the addition of negatively charged membranes. For commercial application an acceptable correlation of results obtained by the enzyme electrode and those of an established laboratory procedure is required. The specificity can be checked by the analysis of the sensitivity to the respective analyte as compared to those of various interfering substances. The optimization of the polymeric material used for immobilization of glucose oxidase resulted in enzyme membranes which are characterized by relative sensitivities of 1:0.6:0.2:0.9 for glucose, ascorbic acid, uric acid, and acetaminophen. Thus, in the presence of 0.2 mmol/l of any of these substances the signal for 5.5 mmol/l glucose in whole blood would not be increased by more than 3.2% for acetaminophen, 2.1% for ascorbic acid, and 0.7% for uric acid (Fig. 1).

Enzyme membranes showing this behaviour (BST Bio Sensor Technologie Berlin) provide a high analytical quality of various enzyme electrode based glucose autoanalyzers (Römer et al., 1990; Bachg and Reinauer, 1995).

The *lifetime* of enzyme electrodes depends on several factors including the specific enzyme activity, the formation of inactivating reaction products, and the operating conditions of the sensor. Thermal protection is neccessary since even matrix-bound enzymes are sensitive to temperatures of more than +40 °C. Inclusion of enzymes like the hydrogen peroxide converting catalase may increase the half-life of hydrogen peroxide sensitive enzymes (Scheller et al., 1989).

Generally, a high enzyme reservoir in the biocatalytic membrane is desirable in order to obtain a high stability. Most commercial enzyme electrodes work with a more than tenfold enzyme excess as compared with that neccessary for a maximum in sensitivity determined by the enzyme loading test (Scheller et al., 1983). This may result in a lifetime up to sever-

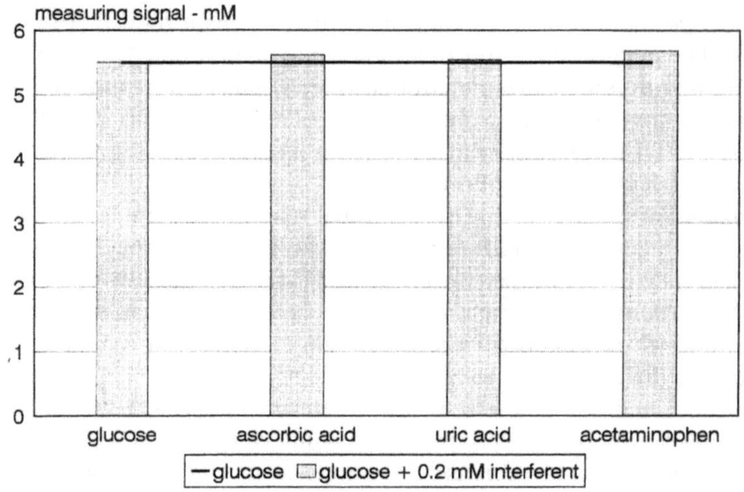

Figure 1. Specificity of glucose oxidase membranes produced by BST Bio Sensor Technologie Berlin.

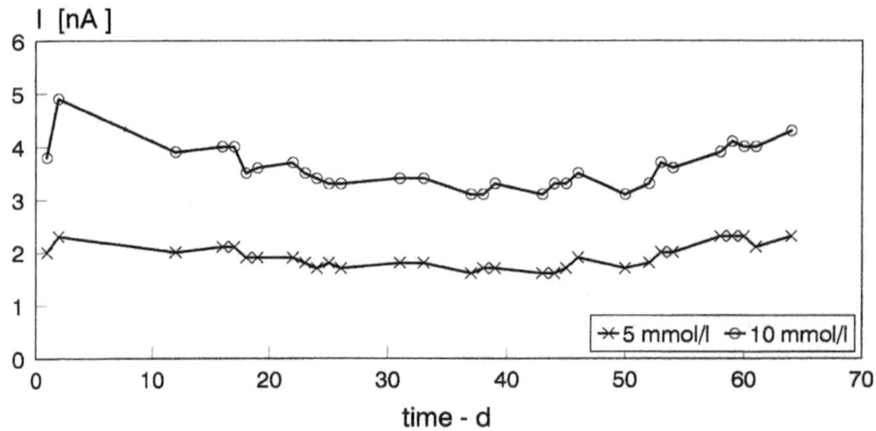

Figure 2. Functional stability of glucose oxidase membrane represented by the time-dependent steady state signal at a working temperature of 37 °C (BST Bio Sensor Technologie).

al months (Fig. 2) and a storage stability of more than 9 months up to 5 years at +4 °C.

Because of the diffusion control at high enzyme activity the glucose sensor signal is not significantly influenced by pH changes between 4.5 and 8.0. However, a further decrease to acidic conditions below pH 4.5 results in an irreversible inactivation of the glucose oxidase (isolated from *Aspergillus niger*) that is completed at pH 3.0 (Fig. 3).

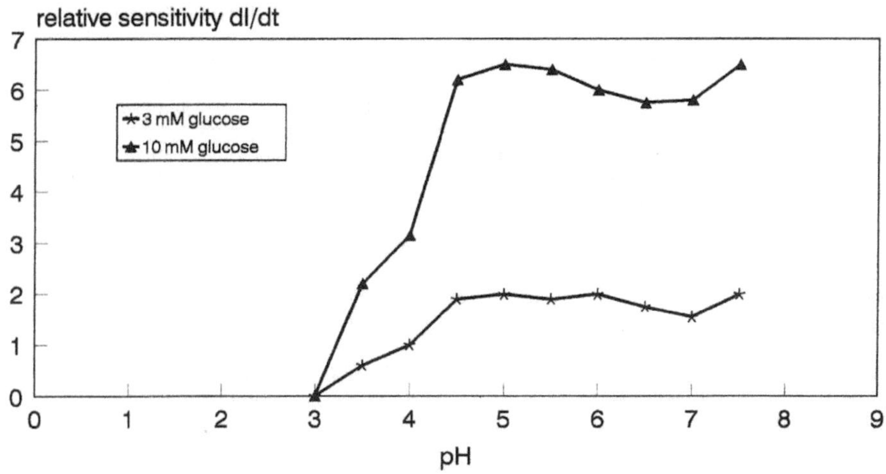

Figure 3. pH dependence of the kinetic signal (dI/dt) of glucose oxidase (*Aspergillus niger*) immobilized in polyurethane; * 0.1 mmol/l phosphate buffer solution, 40 U glucose oxidase/cm^2; * Analysis 15 min after pH change.

Concepts of biosensor-based analytical systems for medical application

Different concepts have been developed to launch biosensors on the market. They are based on the different application fields and include autoanalyzers for *centralized laboratories* working with both prediluted and undiluted samples. *Critical care* demands highly accurate and simple analyzers with a minimum of lagtime between sample withdrawal and result. *On-line monitoring of patients* is offered by the combination on an *ex vivo* detecting unit with microdialysis sampling. To meet the most important challenge to biosensors various groups and companies are concentrating on the *in vivo* analysis by implantable or insertable enzyme sensors. The use of biosensors for diabetic *patient self-monitoring* requires very simple portable systems using unprepared whole blood samples.

Autoanalyzers for centralized laboratories

The first commercial enzyme electrode based analyzer (1975, YSI Corp., USA) was developed to meet the high demand for glucose determination in blood of diabetic patients. Since then, enzyme electrode based analyzers for about 12 different analytes have been commercialized, most of them also for blood glucose analysis, but with increasing importance of other applications. Whereas substances like disaccharides, amino acids, and vitamins are relevant for more or less non-medical fields, glucose, lactate, urea, and uric acid are of interest in clinical chemistry too (Table 1).

Table 1. Stand-alone analyzers based on enzyme electrodes for medical application

Model	Company	Analyte	Measuring Range (mM)	Material	Sample Through-put (1/h)	Func-tion Stability
EBIO 6666	Eppendorf-Netheler-Hinz Hamburg (Germany)	GLUCOSE	0.6–50.0	blood, serum	120	15 d
		LACTATE	0.5–30.0		120	10 d
EBIO *plus*		GLUCOSE	0.6–50.0		180	15 d
		LACTAT	0.5–30.0		180	10 d
EBIO *compact*		GLUCOSE	0.6–50.0		100	15 d
EBIO *classic*		GLUCOSE	0.6–50.0		100	15 d
BIOSEN 5030 L	EKF Industrie Elektronik Magdeburg (Germany)	LACTATE	0.5–30.0	blood, serum	80	10 d
BIOSEN 6030 G		GLUCOSE	0.6–50.0		80	15 d
ESAT 6660	PGW Medingen, Dresden (Germany)	GLUCOSE	0.6–45.0	blood, serum	120	15 d
		LACTATE	0.5–30.0		120	10 d
ESAT 6661		LACTATE	0.5–30.0		120	10 d
		GLUCOSE	0.6–45.0		120	15 d
ECA 2000		GLUCOSE	0.6–45.0		120	15 d
		LACTATE	0.5–30.0		120	10 d
ECA 180		GLUCOSE	0.6–45.0	+ urine	180	15 d
YSI 2300 G	Yellow Springs Instr. (USA)	GLUCOSE	0.0–27.8	blood, plasma	45	7 d
YSI 2300 L		LACTATE	0.0–15.0		45	7 d
Glukometer	Acad. Sciences Berlin (Germany)	GLUCOSE	0.5–50.0	blood,	90	15 d
		LACTATE	0.5–30.0	serum	90	10 d
		URIC ACID	0.1–1.2	serum	40	10 d
GLUCO 20	Fuji Electric Corp.	GLUCOSE	0.0–27.0	blood, serum	90	> 500 samples
UA-300 A	(Japan)	URIC ACID		serum	60	
AUTO-STAT GA 112	Daiichi (Japan)	GLUCOSE	1.0–40.0	blood, serum	120	
EXSAN	Acad. Sci. Lithuania, Inst. Biochem. Vilnius (Lithuania)	GLUCOSE	2.0–30.0	Blood	45	
		LACTATE	0.1–15.0		45	
		UREA	2.0–40.0		30	
		CHOLEST.	0.5–10.0		20	
GLUCO-PROCESSEUR	Tacussel (France)	GLUCOSE	0.05–5.0		90	> 2000 samples

As compared with conventional enzymatic analysis, the main advantages of such analyzers are the extremely low enzyme demand (a few milliunits per sample), the simplicity to operate them, the high analysis speed, and the high analytical quality. The development began with the measurement of discrete samples of about 50 µl whole blood or serum in stirred measuring cells (YSI, Glukometer, Glucoprocesseur, UA-300) with a sample throughput of about 60/h. The respective analyzers are produced for the analysis of glucose, lactate, uric acid, and disaccharides.

An enhancement of sample frequency up to 130/h has been achieved by the integration of segmented continuous flow systems. The resulting one-parameter analyzers have been on the market since the early eighties for metabolites like glucose and lactate. They process prediluted samples and are especially well suited for centralized medical diagnostics. Representatives are ESAT (PGW Medingen, Germany), EBIO (Eppendorf-Netheler-Hinz Hamburg, Germany) (see Fig. 4), BIOSEN 5030L and 5030G (EKF Industrial Electronics Magdeburg, Germany) and Auto-Stat GA-112 (Daiichi, Japan). These analytical systems are characterized by a high analytical quality: The serial imprecision is below 1.5 %, and the results correlate very well with established enzymatic methods (Pfeiffer et al., 1992; Römer et al., 1990; Bachg and Reinauer, 1995). However, in contrast to the conventional methods 10 U of enzyme are enough for the analysis of at least 3000 samples over a period of several months (Fig. 5).

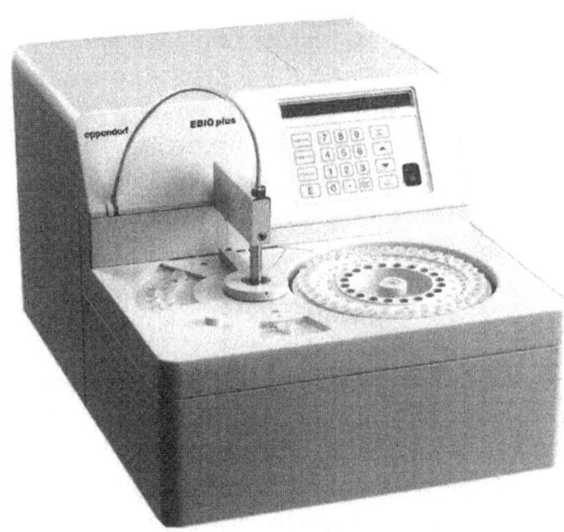

Figure 4. Analyzer EBIO *plus* (Eppendorf-Netheler-Hinz, Hamburg).

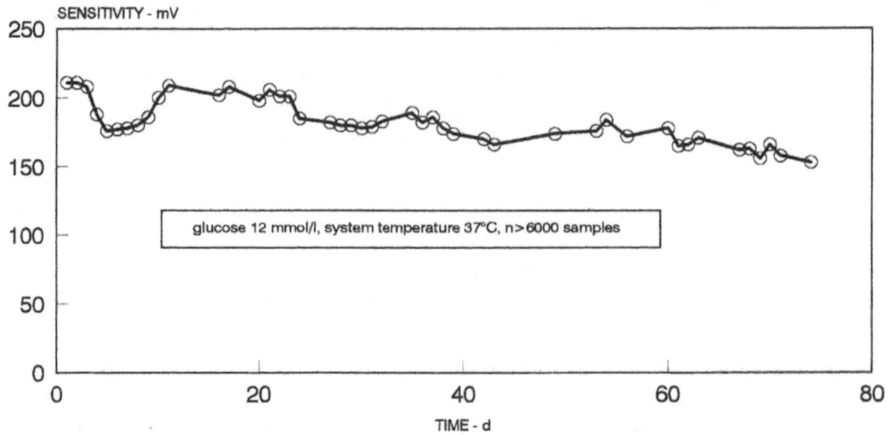

Figure 5. Long-term sensitivity of a glucose sensor in the glucose measuring analyzer EBIO 6660 (Eppendorf-Netheler-Hinz).

Analyzers for pharmaceutical research

A new market segment has been opened up by the successful development of optical immunosensors. Contrary to biospecific electrodes this kind of biosensor is suited to detect binding processes of proteins, polysaccharides and nucleic acids, which is extremely useful in pharmaceutical research to identify direct interactions between high molecular weight substances. The first development in this area by Pharmacia, Sweden, in 1992 (BIAcore, BIAcore 2000, BIAlite) was followed by Fisons Applied Sensor Technlogy, Great Britain (IAsys, 1994), and Artificial Sensing Instruments, Switzerland (GKS1, 1994). However, the technologies behind these devices are rather complicated, and their high price prohibits broad application in medical routine investigations.

Analyzers for critical care

Critical care management basically needs immediate results with high analytical quality. Devices should be portable and able to consolidate the most critical and frequently ordered emergency tests. This may allow for a quick diagnosis of critical metabolic situations and by this way an improved quality of therapy because of timely critical care decisions.

The market for critical care is represented by different types of autoanalyzers (Table 2). They normally measure pH and several electrolytes (K^+, Na^+, Ca^{++}, Cl^-); mostly, blood gases (O_2, CO_2) are available in addition. In 1988 NOVA biomedical Inc. (Waltham, USA) started with the implementation of enzyme sensors for glucose in critical care instruments. Nowadays serveral companies use glucose electrodes and some have succeeded with

Table 2. Critical care monitoring systems with biosensors

Model	Company	Analytes *(biosensor)*	Stability Biosensor
NOVA 7	NOVA biomedical, Waltham (USA)	*lactate,* pH, pCO_2, pO_2	
NOVA 16		*creatinine, glucose,* Na^+, K^+, Cl^-, TCO_2 BUN, Hct	
Serie 800	Ciba-Corning, Medfield (USA)	*glucose, lactate,* pO_2, pCO_2, pH, K^+, Ca^{++}, Na^+, Cl^-	glucose > 3 month
Ionometer EG-Hk	Fresenius AG, Bad Homburg (Germany)	*glucose,* Na^+, K^+, Ca^{++}, pH, Hct, (Hb)	> 2 month
EML 105 ABL 625, 615 or 605	Radiometer, Copenhagen (Denmark)	*glucose,* Na^+, K^+, Ca^{++}, Cl^- *glucose,* Na^+, K^+, Ca^{++}, Cl^-, pO_2, pCO_2	
i-STAT PCA	i-STAT Corp., Princeton (USA)	*glucose, urea,* Na^+, K^+, Cl^-, HcT, (Hb)	disposable

the application of lactate sensors to provide the analysis of the key indicator of oxygen deficiency, infarction, and shock. In 1995 NOVA announced the integration of a creatinine sensor into their 'model 16'.

A significant step forward for critical care diagnosis has been achieved by the commercialization of the portable point-of-care system PCA (i-STAT, Princeton, USA). It is capable of determining electrolytes as well as glucose and urea by a disposable cartridge which is internally calibrated to all analytes just before exposure to the whole blood sample (Jacobs et al., 1993; Erickson and Wilding, 1993). The base transducers are microfabricated thin-film metal sensors for H_2O_2 and ammonium ion sensitive electrodes. The system appears to be highly reliable, however, the high price of the sensors themselves poses a hurdle to large-scale application.

Biosensor-based analytical systems for decentralized facilities

Beside possible influences on the analytical accuracy by undefined patient conditions and sample withdrawal the quality of analysis may be hampered by problems of transport, preanalytics and storage (influence of temperature, time) of the sample. Analysis at the point of care could contribute to the reduction of potential interferences. Frost and Sullivan (1993) mentioned another point considering alternate sites for acute care equipment. Taking into account the desire for a more patient-friendly environment they expect a movement of healthcare out of the hospital leading to an increase in ambulatory healthcare of more than 700% from 1992 to 1998 (finally around an estimated $ 7 billion).

This is a novel challenge for biosensors. Up to now, several disposable enzyme sensors have been brought on the market. The ExacTech and Satellite G (MediSense, USA) and Glucometer Elite (Bayer Diagnostics, Germany) are based on disposable strip electrodes representing enzyme electrodes of the second generation. They use glucose oxidase together with the artificial mediators ferrocene and ferricyanide, respectively. These portable hand-held devices are easy to handle, have low weight, and the prices are low as compared with laboratory analysis. However, it has to be considered that the level of information and reliability provided by a glucose control by one-way strip electrodes is not comparable with analysis made in a medical laboratory. One-way procedures are useful for an additional analysis at home, but do not substitute the control and monitoring by a medical doctor. The basic disadvantage of strip electrodes seems to be that there is no possibility to calibrate the electrode – it is applicable just for one analysis. This has been changed by the i-STAT device (see above), useful for the analysis of glucose and urea besides electrolytes where an internal calibration of one-way cartridges has been included.

On the other hand, the portable enzyme membrane electrode based systems BIOSEN 5020L and 6020G (EKF Industrie Elektronik) work with reusable enzyme membranes and internal calibration. One membrane containing 10 U of glucose oxidase is stable for more than 30 days. During this period more than 3000 analyses can be performed. Beside simple handling and an extremely short lag time between sample withdrawal and measuring result these devices are characterized by an analytical performance comparable with that of clinical laboratories. Therefore, and based on their mobility, these systems are particularly well suited for application the different sports facilities and intensive care units separate from centralized laboratories. Because of the enormous lifetime of 10 U of enzyme and the low contribution of additional consumables the costs per sample are reduced by 35% in case of 20 samples per day, by 50% in case 30 samples per day and about 70% in case of 50 samples per day as compared with the price of 1.00 DM per test strip.

Real-time analysis

The real-time analysis of substances, including metabolites and enzymes, is one of the most profound challenges to biosensors. However, direct analysis in humans has a critical demand for hemocompatibility and must not cause any cytotoxic, carcinogenic or genotoxic reactions, inflammation, irritation or sensitization. In dependence on the category of the respective medical device, the body contact, and the contact duration, several biological effects have to be investigated by competent notified bodies (Bünger and Tümmler, 1994). Although optimistic results concerning invasive glucose analysis have been obtained by various groups from Neuchâtel, Switzerland (Koudelka et al., 1993), Vienna, Austria (Urban et al., 1992), Kansas, USA-Paris, France (Wilson et al., 1992), San Diego, USA (Armour et al., 1990) and Karlsburg, Germany (v. Woedtke et al., 1991) the problems mentioned are responsible for the fact that no truely reliable implantable biosensor is available on the market today.

In the meantime sampling by microdialysis combined with an *ex vivo* detection of target analyte concentration seems to be helpful. Problems in obtaining a defined recovery of the dialysis that is dependent on temperature, flow, nature of the sample and the condition of the dialysis membrane, may be compensated by external calibration. On-line glucose monitoring systems based on microdialysis sampling have been brought on the market by unitec Ulm, Germany ('Glucosensor'), and Ampliscientifica Milano, Italy ('Glucoday'). These systems have been successfully applied to the monitoring of diabetics over several hours, resulting in the improvement of the assessment of labile patients.

With the exception of critical care instruments the commercial biosensor-based devices today are stand-alone analyzers. This concept will probably change in the near future. Biosensors should offer the possibility to analyze whole panels of substances with respect to special fields of interest within a short time and with high accuracy close to the location of sample withdrawal. This will be one way for biosensors to be established in clinical use. Another way could be an increased investment for invasive application. Any successful commercial development in this direction will cause a general improvement of the value of biosensorics.

References

Armour, J.C., Lucisano, J.Y., McKean, B.D. and Gough, D.A. (1990) Application of chronic intravascular blood glucose sensor in dogs. *Diabetes* 39:1519–1526.

Bachg, D. and Reinauer, H. (1995) Der economic Glucoseanalyser – Evaluationsergebnisse. *Lab. med.* 19:463–466.

Bünger, M. and Tümmler, H.-P. (1994) Classification of active medical devices according to the medical device directive. *Medical Device Technology* 5/2:33–39.

Erickson, K.A. and Wilding, P. (1993) Evaluation of a novel point-of-care system, the i-STAT portable clinical analyzer. *Clin. Chem.* 39/2:283–287.

Frost and Sullivan (1994) *Report on alternate site acute care equipment and supply markets.* Mt. View, USA.

Hasebe, Y., Terata, S. and Uchiyama, S. (1993) Chemically amplified adrenal medulla hormon sensor based on subatrate recycling using tyrosinase and L-ascorbic acid. *Analytical Sciences* 9:855–861.

Jacobs, E., Vadasdi, E., Sarkozi, L. and Colman, N. (1993) Analytical evaluation of i-STAT portable clinical analyzer and use by nonlaboratory health-care professionals. *Clin. Chem.* 39/6:1069–1074.

Koudelka-Hep, M., Strike, D. and de Rooij, N.F. (1993) Miniature electrochemical glucose biosensors. *Anal. Chim. Acta* 281:461–466.

Mascini, M. and Vadgama, P. (1993) Chemo- and biosensors for medical diagnosis. *In:* A.P.F. Turner (Ed.) Advances in biosensors. Supplement 1. JAI Press Ltd, pp. 109–139.

Mizutani, F., Yabuki, S. and Asai, M. (1991) Highly sensitive measurement of hydroquinone with an enzyme electrode. *Biosens. Bioelectron.* 6:305–310.

Olsson, B., Lundbäck, H., Johansson, G., Scheller, F. and Nentwig, J. (1986) Theory and application of diffusion-limited amperometric enzyme electrode detection in flow injection analysis of glucose. *Anal. Chem.* 58:1046–1052.

Pfeiffer, D., Scheller, F., Jänchen, M. and Bertermann, K. (1980) Glucose oxidase bienzyme electrodes for ATP, NAD^+, starch and disaccharides. *Biochimie* 62:587–593.

Pfeiffer, D., Setz, K., Schulmeister, T., Scheller, F.W., Lück, H.-B. and Pfeiffer, D. (1992) Development and characterization of an enzyme-based lactate probe for undiluted media. *Biosens. Bioelectron.* 7:661–671.

Pfeiffer, D., Scheller, F.W., Setz, K. and Schubert, F. (1993) Amperometric enzyme electrodes for lactate and glucose determination in highly diluted and undiluted media. *Anal. Chim. Acta* 281:489–502.

Römer, M., Haeckel, R., Bonini, P., Ceriotti, G., Vassault, A., Solere, P. and Morer, P. (1990) European Multicentre Evaluation of the ESAT 6660. *J. Clin. Chem. Clin. Biochem.* 28: 435–443.

Scheller, F., Seyer, I., Scheller, O., Gesierich, A., Deutsch, K., Makower, A. and Jänchen, M. (1978) DD patent 131 414.

Scheller, F., Pfeiffer, D., Seyer, I., Kirstein, D., Schulmeister, T. and Nentwig, J. (1983) Degree of glucose conversion in amperometric glucose oxidase -mutarotase membrane electrodes. *Bioelectrochem. Bioenerg.* 11:155–165.

Scheller, F., Pfeiffer, D., Hintsche, R., Dransfeld, I. and Nentwig, J. (1989) Glucose measurement in diluted blood. *Biomed. Biochim. Acta* 48:891–896.

Scheller, F., Makower, A., Ghindilis, A., Bier, F., Förster, E., Wollenberger, U., Bauer, Ch., Micheel, B., Pfeiffer, D., Szeponik, J., Michael, N. and Kaden, H. (1995) Enzyme sensors for subnanomolar concentrations. *In*: K.R. Rogers, A. Mulchandani and W. Zhou (eds): *Biosensor and chemical sensor technology. Process Monitoring and control.* ACS Symposium series 613. American Chemical Society, Washington, DC, pp 70–81.

Schubert, F., Kirstein, D., Schröder, L. and Scheller, F. (1985) Enzyme electrodes with substrate and co-enzyme amplification. *Analyt. Chim. Acta* 169:391–396.

Urban, G., Jobst, G., Keplinger, F., Aschauer, E., Tilado, O., Fasching, R. and Kohl, F. (1992) Miniaturized multi-enzyme biosensors integrated with pH sensors on flexible polymer carriers for *in vivo* applications. *Biosens. Bioelectron.* 7:7–13.

Wasa, T., Akimoto, K., Yao, T. and Murao, S. (1984) Development of laccase membrane electrode by using carbon electrode impregnated with epoxy resin and its response characteristics. *Nippon Kagaku Kaishi.* 9:1397–1398.

Wilson, G., Zhang, Y., Reach, G., Moatti-Sirat, D., Poitout, V., Thevenot, D.R., Lemonnier, F. and Klein, J.-C. (1992) Progress toward the development of implantable sensor for glucose. *Clin. Chem.* 38:1613–1617.

von Woedtke, T., Fischer, U., Brunstein, E., Rebrin, K. and Abel, P. (1991) Implantable glucose sensors: Comparison between *in vitro* and *in vivo* kinetics. *Int. J. Artif. Organs* 14:473–481.

Wollenberger, U., Schubert, F., Pfeiffer, D. and Scheller, F. (1993) Enhancing biosensor performance using multienzyme systems. *TIBTECH.* 11:255–262.

Yang, X., Pfeiffer, D., Johansson, G. and Scheller, F. (1991) Nanomolar level amperometric determination of ATP through substrate recycling in an enzyme reactor in a FIA system. *Anal. Letters* 24/8:1401–1417.

Frontiers in Biosensorics II
Practical Applications
ed. by. F.W. Scheller, F. Schubert and J. Fedrowitz

Biosensors with modified electrodes for *in vivo* and *ex vivo* applications

G. A. Urban and G. Jobst

Institut für allgemeine Elektrotechnik und Elektronik, L. Boltzmann-Institut für biomedizinische Mikrotechnik und Hirnkreislaufforschung, Technical University Vienna, 1040 Wien, Austria

Summary. Integrated and miniaturized biosensor arrays were developed exhibiting outstanding performance. Biosensors with negligible sensitivity to interferences and high long-term stability were produced by modifying electrochemical transducers and utilizing photopatternable enzyme membranes. The use of appropriate miniaturization technology leads to mass producible devices for *in vivo* and *ex vivo* applications.

Introduction

There is a strong demand, particularly in diabetology, in the intensive care unit, the operation theater and in the field of bed side analysis for miniaturized integrated biosensors *in vitro, ex vivo* and *in vivo* as reported by Astrup et al. (1977) and Strang et al. (1973).

Most of the numerous biosensing devices reported in the literature suffer from the serious drawback that the sensor performance is rather poor in respect to interferences and long-term stability. Additionally the production processes for conventional biosensor devices are rather delicate and therefore hard to automate. This fact not only increases production costs but can also cause serious problems with the reliability of such devices. Therefore much effort is concerned with the development of biosensors made by means of established mass production technologies. Screen printing and thin film technology are most expected to successfully face this challenge. Screen printing technology has already proved its ability to manufacture reliable disposable biosensors (Matthews et al., 1987).

Furthermore, to create integrated biosensors for measurements of the different metabolic parameters required in medicine (Astrup et al., 1977; Strang et al., 1973), the employed technology has to enable the selective immobilization of different enzymes on their corresponding electrodes. Different approaches for the regioselective deposition of functional membranes have been reported including drop-on techniques (Hintsche et al., 1990), ink-jet printing (Kimura et al., 1988), spray techniques (Yokoyama et al., 1989), electropolymerization (Tamiya et al., 1989), lift-off techniques (Murakami et al., 1986), screen-printing technology (Bilitewski, 1992), enzyme membranes deposited by electrodeposition (Mastrototaro et al., 1991), and photolithographically patterned enzyme membranes (Shiono et al., 1986; Hanazato et al., 1989; Wilke and Müller, 1992).

A straightforward approach is the use of membranes that can be directly patterned by photolithography. Advantageous is the UV-initiated free radical cross linking of the polymer directly on the substrate which allows to design membranes with different physico-chemical properties simply by altering the composition of the UV-sensitive membrane precursor or UV exposure time. Up to now, however such attempts provided a low measuring range (Shiono et al., 1986; Wilke and Müller 1992) and in the case of integrated multi-enzyme sensors, they suffered from the danger of enzyme mixing.

A thin film process is presented in this paper which overcomes the drawbacks by using modified electrodes and photopatternable membranes.

Experimental section

Apparatus

The electrochemical measuring setup consisted of a home-made SMD bipotentiostat linked to a PC-based data acquisition and actuating program written in Asyst 4.0 (Asyst Technologies Inc.) via an PCL818 ADC board (Advantech). Electrochemical characterization was performed in a flow-through cell with 1 mm^2 diameter of the flow channel or by simply inserting the sensor into a magnetically stirred solution in a beaker.

Reagents

Upilex^R substrates were from ICI (Vienna), Ti, Pt and Ag were obtained from Balzers (Liechtenstein), the enzymes glucose oxidase GOD (EC 1.1.3.4, aspergillus niger, Biozyme GO3A, 360 U/mg protein) and catalase (EC 1.11.1.6, aspergillus niger, Biozyme CATANIF, 3000 U/mg protein) were kindly provided by Biozyme UK, lactate oxidase LOD (EC 1.1.3.2, aerococcus viridans, Asahi chem. ltd. LOD II) was kindly provided by Genzyme UK. Polyimide photo resists are a product of OCG Switzerland. Irgacure 651 (ω,ω'-dimethoxy ω-phenyl acetophenone) was kindly provided by Ciba Geigy. Hydroxyethyl methacrylate (HEMA) (> 95%) and tetraethylene glycol dimethacrylate (TEGDMA) (75–85%) comes from Fluka, pHEMA was obtained from Polyscience, all other chemicals were of p. a. grade.

A typical hydrogel precursor consists of 28% pHEMA as polymeric binder, 28% HEMA as reactive monomer, 3% TEGDMA as cross linker, 40% ethylene glycol as plasticizer, and 1% Irgacure 651 as photo initiator. After dissolution of all compounds it gave a clear colorless solution which was finally 0.2 μm filtrated. To this precursor solution the desired enzymes were added to give the enzyme membrane precursors containing up to 5 wt% of the proteins.

Device preparation

A thin film process is presented in this paper which overcomes the mentioned drawbacks by immobilizing the enzymes glucose oxidase, lactate oxidase, and catalase into pHEMA-hydrogel membranes that were structured by photolithography. Separated by an enzyme-free membrane a catalase membrane was created topmost on this multi-membrane setup in order to prevent crosstalking and the release of hydrogen peroxide to the analyte (Urban et al., 1992). A miniaturized flexible carrier was chosen to prevent tissue damage in case such a sensor was intended for *in vivo* measurements. To suppress electrochemical interferences an electropolymerized semipermeable membrane was introduced. Figure 1 gives a schematic view of the sensor construction and the main reaction and transport pathways (Urban et al., 1992). Tests were performed in buffer solution and serum to characterize the devices.

In-vivo *biosensor device*

Sixty biosensor devices on one wafer, 60 mm long and 0.7 mm wide, comprising two platinum working electrodes of 0.4 mm^2 area, one platinum counter electrode and an Ag/AgCl pseudo reference electrode insulated by

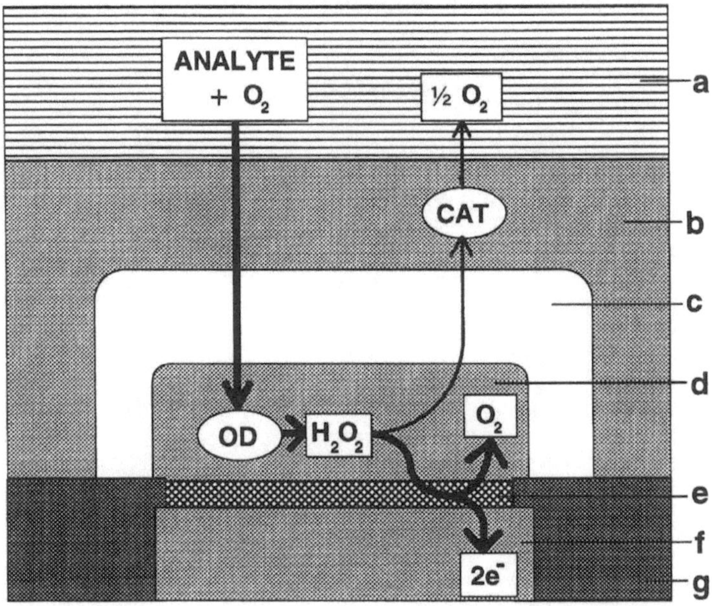

Figure 1. Schematic drawing of sensor build-up and the dominant reaction and transport pathways. a = bulk solution, b = catalase membrane, c = spacer membrane, d = oxidase membrane, e = electropolymerized permselective membrane, f = platinum anode, g = insulation.

a 1-μm thick photo patternable polyimide dielectric are produced on top of a 0.1-mm thick highly flexible polymer carrier (Upilex®). The semipermeable membrane was formed by electropolymerization of 1,3-diamino benzene in aqueus solution according to the procedure described by Gleise et al. (1991). The electropolymerization was performed for at least three hours on the wafer stage with all working electrodes electrically interconnected. The membrane precursors were applied to the wafers by a standard spin-on technique, the resulting films were exposed to UV-light through a photo mask under argon flushing. The exposed layers were subsequently developed in ethylene glycol/water 1:1 (wt/wt). The multimembrane arrangement was realized by repetition of this basic process with different photo masks and membrane precursors (Urban et al., 1992). Figure 2 shows a schematic drawing of the integrated device.

Ex-vivo *biosensor device*

The assembling of a thin film sensor array consisting of four working electrodes and one reference electrode with a printed circuit board performing both electrical and fluidic functions gives an analytical micro flow system (Fig. 3) with a flow cell volume of 1.5 μl. The area of the working electrodes is 0.5×0.5 mm^2 each and described in detail in Urban et al. (1994b).

Since this device is made from two components made by means of well-established mass production technologies, thin film and printed circuit

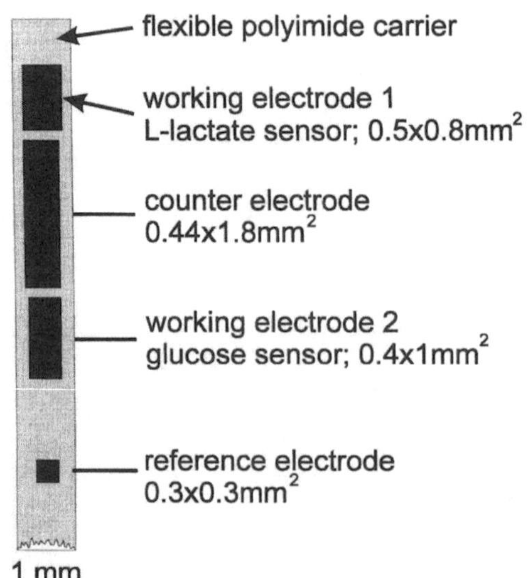

Figure 2. Schematic drawing of the integrated *in vivo* glucose/lactate device (conducting lines are not shown).

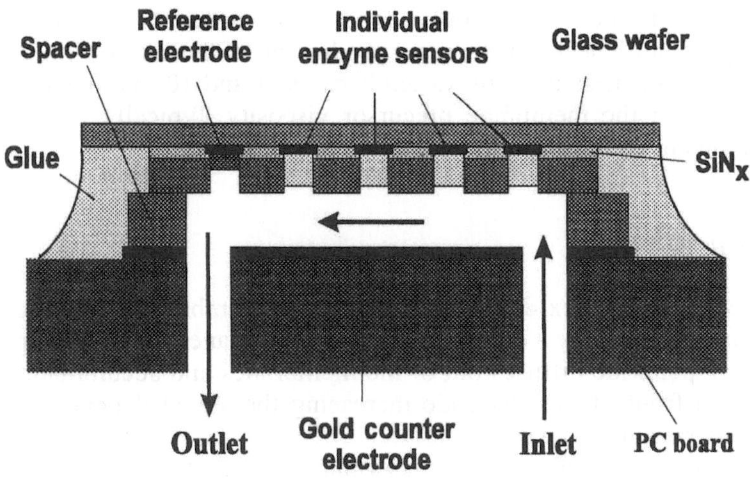

Figure 3. Schematic cross section of a microflow device with assembled biosensor array.

board technology, and assembling of the parts is compatible to IC packaging techniques, mass fabrication of this device is possible with acceptable effort.

Electrochemical measurements

Electrochemical oxidation of hydrogen peroxide was employed as transducing principle. Oxidation of H_2O_2 was performed on both Pt-working electrodes at +600 mV versus the internal Ag/AgCl pseudo reference electrode. Measurements were carried out in 0.1 M phosphate buffer with 0.1 M NaCl (pH 7.4) and undiluted bovine serum. Different glucose and lactate concentrations were realized by addition of the desired amount of 1 molar glucose or L-lactate stock solutions, respectively.

Results and discussion

Fabrication technology

pHEMA is advantageous as membrane material because of its well-known blood compatibility (Margules et al., 1983; Jeyanthi and Rao Panduranga, 1990), and its high mechanical strength in the swollen state. Additionally photopatternable precursor solutions can be made with enzyme compatible solvent mixtures. Furthermore, by variation of precursor composition and cross linking conditions physicochemical membrane properties can be varied over a broad range. The handling procedure of the enzymatic and non enzymatic hydrogel precursors are compatible to a large extent with common thin film technology processes. Electropolymerization of the semipermeable membrane can be done on the wafer stage and therefore presents no bottleneck for the mass production of these thin film devices. Membrane thickness is easily varied between 1 and 10 microns by proper adjustment of the membrane precursor viscosity. Typically, membranes with a thickness of 4–6 μm are employed in this work.

Device performance

With conventional oxidase-based biosensor membrane setups, oxidase membrane covered by a diffusion limiting membrane, the reaction product hydrogen peroxide diffuses out of the membranes and accumulates in the analyte in front of the electrode increasing the signal depending on the exchange rate of the analyte.

Simultaneously, substrate and oxygen are depleted from the analyte decreasing the signal. Therefore sensor reading is influenced by the flow speed of the analyte.

One way to minimize this problem is to eliminate the hydrogen peroxide accumulation by topcoating the device with a catalase membrane and keep the analyte consumption low by proper diffusion limitation.

The flow sensitivity of a glucose/lactate device with such a three-membrane build-up is shown in Figure 4. Since the measurement was done in a flow channel with an area of 1 mm^2 the flow rates indicated correspond to flow velocities of mm/min. The current of both sensors is decreased by only 3% when the flow rate is reduced three orders of magnitude from 240 μl/min to 0.24 μl/min, which demonstrates that no hydrogen peroxide is released to the analyte. This behavior seems to be very useful for subcutaneous sensing since the tissue reaction to the cytotoxic agent hydrogen peroxide (Zhang, 1991) is not fully evaluated yet and one can expect that it is not useful for the reliability of *in vivo* measurement.

The calibration graph for this device is shown in Figure 5. The sensitivities of the biosensors can be varied over wide ranges (2–10 nA nM/mm^2 glucose and 5–25 nA nM/mm^2 L-lactate) by proper choice of the thickness of the membranes and by alteration of the hydrogel precursor composition. Relative standard deviations fo the sensitivities of flexible devices from the same batch were measured to be 3.1% ($n = 10$) for the glucose sensors and 3.2% ($n = 10$) for the lactate sensors. Within run reproducibility was tested for glucose devices in bovine serum spiked with glucose, and standard deviations of 0.38% ($n = 10$) at 5 mmol glucose/l and 0.27% ($n = 10$) at 33.4 mmol glucose/l were obtained (Urban et al., 1994a). The base current decreased in less than one hour to below 1 nA/mm^2 with the first use of the device. Hydration of the device proceeds very fast due to the thin membranes. Devices with linear measuring ranges of 25 mmol L-lactate/l and 40 mmol glucose/l display 95% response times of 15 and 25 seconds for changes in L-lactate and glucose concentrations, respectively.

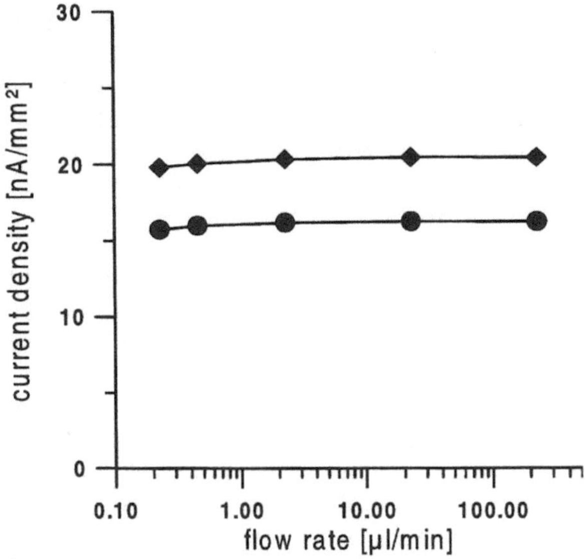

Figure 4. Flow sensitivity of an integrated glucose/lactate device. ◆ L-lactate sensor, ● glucose sensor. The measurement was done in phosphate buffered saline containing 5 mmol/l glucose and 2 mmol/l L-lactate.

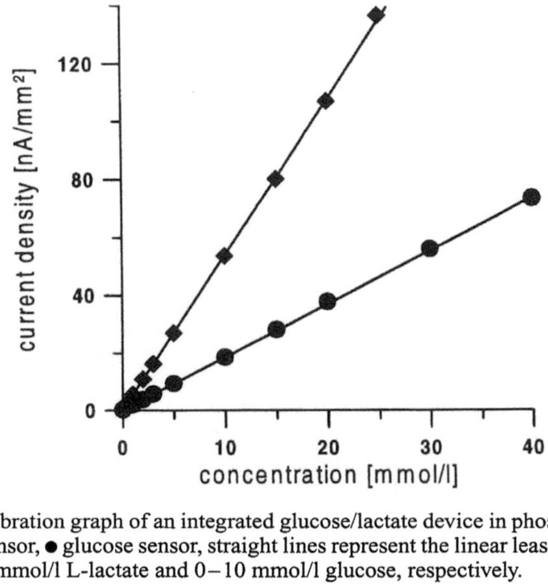

Figure 5. Calibration graph of an integrated glucose/lactate device in phoshate buffered saline. ◆ L-lactate sensor, ● glucose sensor, straight lines represent the linear least squares fit of values between 0–5 mmol/l L-lactate and 0–10 mmol/l glucose, respectively.

The influence of interferences was tested with a device which had already been continuously operated for two days in undiluted bovine serum. The concentrations of the interferences tested reflect the upper physiological level. Only paracetamol at the toxic level of 2.0 mmol/l gave a pronounced increase in sensor reading. Nevertheless the error in reading is less than 0.5 mmol/l glucose and 0.2 mmol/l L-lactate respectively due to the modification of the electrodes using semipermeable layers.

The temperature coefficient of both the glucose and the lactate sensors is 2.5%/°C. There is also no effect of hematocrite (Urban et al., 1992) on the sensitivity of the device. Operational stability of the flexible integrated glucose/lactate device when operated continuously in undiluted bovine serum is more than two weeks at ambient temperature.

The laboratory on chip for *ex vivo* aplications shown in Figure 3 was evaluated during diabetological provocation tests on human volunteers using undiluted heparinized venous blood as analyte. Intravenous glucose tolerance tests (IVGTT) and oral glucose tolerance tests (OGTT) were performed. Venous blood was continuously withdrawn and heparinized over a period of six hours by means of a double lumen catheter. Blood samples were taken manually in order to get samples for glucose and lactate analysis in the clinical laboratory.

Sensor currents measured were converted to concentrations by application of the sensitivities as determined before the experiment with a one-point calibration in buffer solution. Figures 6 and 7 show the time course of glucose and lactate levels of a combined IVGTT/OGTT experiment. The good agreement between sensor and laboratory results prove the effec-

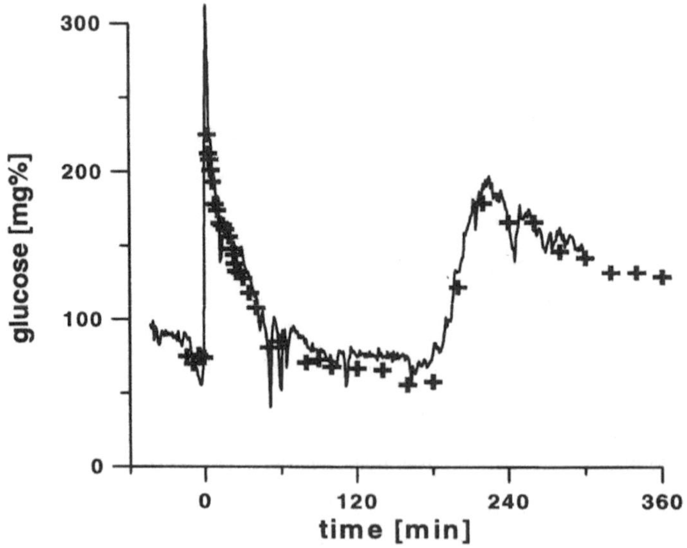

Figure 6. Glucose sensor reading (solid line) and reference values of an IVGTT/OGTT.

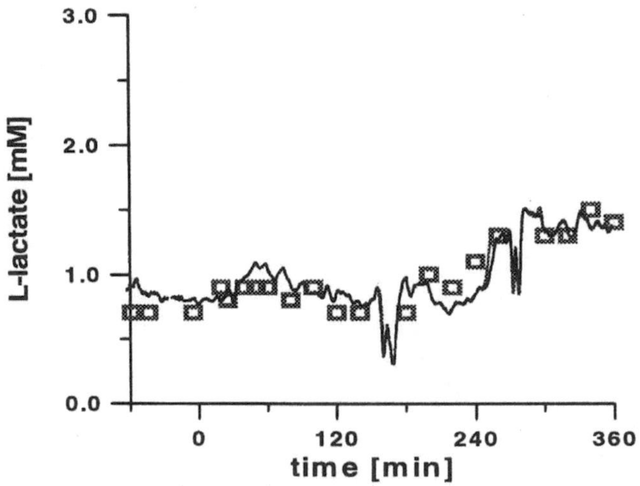

Figure 7. L-lactate sensor reading (solid line) and reference values of an IVGTT/OGTT.

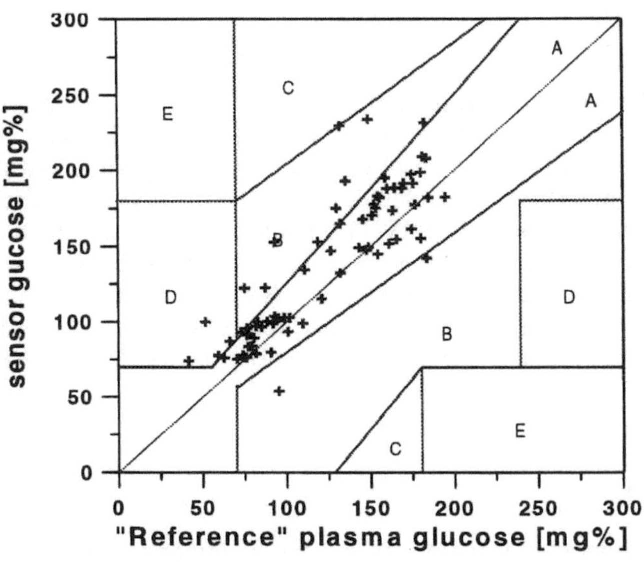

Figure 8. Error grid analysis of four IVGTT's.

tiveness of the precalibration procedure. Sensitivities determined after the experiments deviated less then 10% from the pre-calibration values.

Figures 8 and 9 show the correlated sensor/laboratory results in an error grid analysis. This error grid analysis reflects the clinical demands on the accuracy of glucose measurements. Zone A is related to "clinically accurate", B to "clinically acceptable" and the others to "not acceptable". In

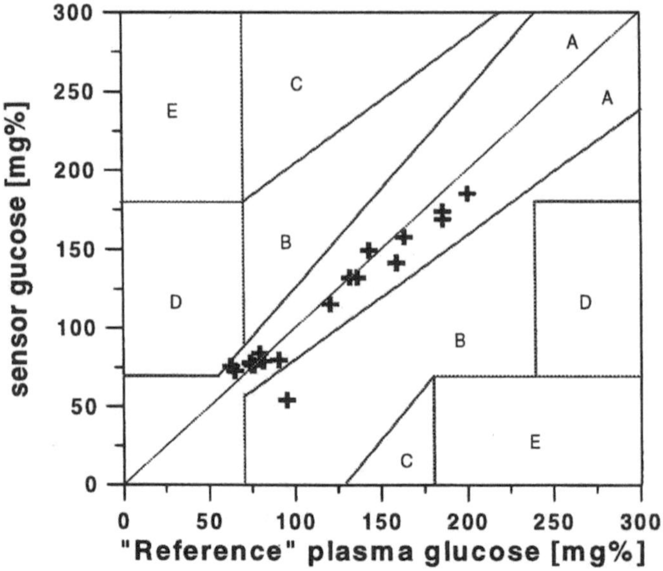

Figure 9. Error grid analysis of four OGTT's.

Figure 8, which holds the results of the IVGTT period of four experiments, 79% of the values are in zone A, 16% in B, and 5% in C, D and E. In contrast, the OGTT period results (of the same experiments) shown in Figure 9, are much closer correlated. Zone A holds 93% of the data points. The poorer correlation of the IVGTT results as compared to the OGTT results can be explained by the very fast concentration changes which can not be detected in time with *in vitro* methods caused by intravenous glucose administration. In the OGTT periods, which reflect normal physiological conditions, the miniaturized mass producible device fulfills the clinical demands.

Conclusions

The results show that different sensors can be integrated to give miniaturized multisensing devices. The modification of the electrode reduces the sensitivity against interferences, and photopatternable enzyme membranes yield integrated biosensing devices with high quality. Furthermore, the performance also fits the needs of clinical analyzer applications. The possibility to miniaturize these devices allows to run an analyzer with very small sample and reagent volumes. This is one major step towards the pocket-sized clinical analyzer and continuous operating *ex vivo* analyzers. Microsystem technology leads to an *ex vivo* device which can be operated in a real¯clinical environment.

Additionally, the flexible polymer carrier and the outstanding sensor performance make the biosensors well suited for subcutaneous measurement.

Clinical monitoring by means of integrated biosensors seems feasible in the next future.

Acknowledgement
This work was supported by the Fonds zur Förderung der Wissenschaftlichen Forschung, the Forschungsförderungsfonds für die Gewerbliche Wirtschaft and the society for microelectronics (GME).

References

Astrup, J., Symon, L., Branston, N. and Lassen, N. (1977) *Stroke* 8:51.
Bilitewski, U. (1992) *In*: G. Costa and S. Miertus (eds): *Trends in electrochemical Biosensors* World Scientific, Singapore, pp 59–68.
Gleise, R., Adams, J.M., Barone, N.J. and Yacynych, A.M. (1991) *Biosens. Bioelectron.* 6:151–160.
Hanazato, Y., Nakako, M., Satorus, S. and Mitsuo, M. (1989) *IEEE Trans. Electron. Dev.* 36:7, 1303–1310.
Hintsche, R., Dransfeld, I., Scheller, F., Pham, M., Hoffmann, W., Hueller, J. and Moritz, W. (1990) *Biosens. Bioelectron.* 5:327–334.
Jeyanthi, R. and Rao Panduranga, R. (1990) *Biomaterials* 11:238–243.
Kimura, J., Kawana, Y. and Kuriyama, T. (1988) *Biosensors* 4:41–52.
Margules, G.S., Hunter, C.M. and MacGregor, D.C. (1983) *Med. Biol. Eng. Comput.* 21:1–8.
Mastrototaro, J., Johnson, K., Morff, R., Lipson, D., Andrew, C. Allen, D. (1991) *Sensor. Actuator. B* 1–4:139–145.
Matthews, D., Holman, R., Streemson, J., Watson, A. and Hughes, S. (1987) *Lancet* 778.
Murakami, T., Nakamoto, S., Kimura, J., Kuriyama, T. and Karube, I. (1986) *Anal. Lett.* 19:1973–1986.
Shiono, S., Hanazato, Y. and Nakako, M. (1986) *Anal. Sci.* 2:517–521.
Strang, R.H.C. and Bachelard, H.S. (1973) *J. Neurochem.* 20:987.
Tamiya, E., Karube, I., Hattori, S., Suzuki, M. and Yokoyama, K. (1989) *Sensor. Actuator. B* 18:297–307.
Urban, G., Jobst, G., Keplinger, F., Aschauer, E., Tilado, O., Fasching, R. and Kohl, F. (1992) *Biosens. Bioelectron.* 7:733–739.
Urban, G., Jobst, G., Aschauer, E., Tilado, O., Svasek, P. and Varahram, M. (1994a) Performance of integrated glucose and lectate thin film microbiosensors for clinical analyzers. *Sensor. Actuator. B* 1–3:592–596.
Urban, G., Jobst, G., Svasek, P., Varahram, M., Moser, I. and Aschauer, E. (1994b) Development of a micro flow-system with integrated biosensor array. *MESA Monographs, Micro Total Analysis System, Univ, Twente*, p 249.
Wilke, D. and Müller, H. (1992) *In*: G. Costa and S. Miertus (eds) *Trends in electrochemical Biosensors*. World Scientific: Singapore, pp 155–162.
Yokoyama, K., Sode, K., Tamiya, E. and Karube, I. (1989) *Anal. Chim. Acta* 218:137–142.
Zhang, Y, Bindra, D.S., Barrau, M.B. and Wilson, S. (1991) *Biosens. Bioelectron.* 6:653–661.

Subject Index

MIX
Papier aus verantwortungsvollen Quellen
Paper from responsible sources
FSC® C105338

If you have any concerns about our products,
you can contact us on
ProductSafety@springernature.com

In case Publisher is established outside the EU,
the EU authorized representative is:
Springer Nature Customer Service Center GmbH
Europaplatz 3, 69115 Heidelberg, Germany

Printed by Libri Plureos GmbH
in Hamburg, Germany